Our Biggest Fight

Our Biggest Fight

Reclaiming Liberty, Humanity, and Dignity in the Digital Age

Frank H. McCourt, Jr.
with Michael J. Casey

CROWN
NEW YORK

Published in the United States by Crown, an imprint of the Crown Publishing Group, a division of Penguin Random House LLC, New York.
crownpublishing.com

Crown and the Crown colophon are registered trademarks of Penguin Random House LLC.

Library of Congress Cataloging-in-Publication Data
Names: McCourt, Frank H., Jr., author. | Casey, Michael, 1967– author.
Title: Our biggest fight : reclaiming liberty, humanity, and dignity in the digital age / by Frank H. McCourt, Jr., with Michael J. Casey.
Identifiers: LCCN 2023056212 | ISBN 9780593728512 (hardback ; alk. paper) | ISBN 9780593728536 (trade paperback ; alk. paper) | ISBN 9780593728529 (ebook)
Subjects: LCSH: Internet—Social aspects. | Information technology—Social aspects.
Classification: LCC HM851 .M3936 2024 | DDC 303.48/33—dc23/eng/20240110
LC record available at https://lccn.loc.gov/2023056212

ISBN 978-0-593-72851-2
Ebook ISBN 978-0-593-72852-9

Printed in the United States of America on acid-free paper

Editor: Paul Whitlatch
Editorial assistant: Katie Berry
Production editor: Joyce Wong
Text designer: Andrea Lau
Production manager: Heather Williamson
Managing editor: Christine Tanigawa
Copy editor: Janet Renard
Proofreader: Ethan Campbell

9 8 7 6 5 4 3 2 1

1st Printing

First Edition

Book design by Andrea Lau

To my mother, Kay McCourt, who for over
a century never lost her optimism about the future.
—FHM

To Phillip, Bods, and Des. It's all about "the challenges."
—MJC

Start by doing what's necessary. Then do what's possible, and suddenly you are doing the impossible.
—Saint Francis of Assisi

Raise your hopeful voice.
You have a choice.
You'll make it now.
—Glen Hansard and Markéta Irglová, "Falling Slowly"

CONTENTS

Our Biggest Fight

INTRODUCTION

Hidden in Plain Sight

We face a choice about our future.

Do we want to envision, write, and be in charge of a future in which we are respected as individuals and in which we can enhance and enrich our society? Or do we want our future to be written by a few giant corporations whose technology, algorithms, and devices steadily chip away at our humanity? It's a choice between human beings and machines. Everything—and I mean *everything*—hinges on this decision.

The future will be digital. There is no turning back the information technology revolution that has transformed our lives and blurred the lines between our digital and physical existences. The internet, the most profound expression of that digital transformation, will necessarily remain integral to our economic, social, and political organizations. Still, the future I would choose is one in which the internet has a far more positive impact on our lives than it does today. It's a digital future that has humans at its center.

The internet began as a utopian dream. Well, as we'll learn, the idea was first shaped, at least partly, by military considerations. But the internet of popular imagination, the one that some visionary academics, Silicon Valley's big-thinking coders, and a community of cyber-hippies latched on to in the 1980s and '90s, the internet of mass disruption and freedom of information—*that* idea of an internet was one that was going to liberate us. Powerful elites would no longer control information systems or dominate the commercial networks that defined the economy. The "marketplace of ideas" would organically bubble up the best of humanity and give scientists and innovators, wherever they might be, a real chance to improve life on this planet. It would be an internet of freedom, of unfettered access, of inclusion, of diversity, of collaborative and constructive connections between all people everywhere.

That's not the internet we ended up with.

As it currently functions, the internet—despite all the conveniences and connectivity it has unleashed—is the primary cause of a pervasive unease in the United States and other democratic societies. The internet explains why our national arguments seem intractable. It's why every issue is reduced, in public debate, to the lowest common denominator. It's why youth suicide rates are rising, why politics is filled with toxicity, and why many now dread Thanksgiving dinners. It's why some people in California, much like their red-state counterparts in Texas, are seriously contemplating secession. It's why armed guards are deployed at school board meetings. It is why a massive law enforcement effort has failed to curtail the spread of a deadly fentanyl epidemic, with supply routes of the drug secretly coordinated over social media. It's why Americans' confidence in a variety of civil institutions, as measured by the Edelman Trust Barometer, is at an all-time low. It's why people are calling this a post-truth society.

On the face of it, these issues might seem unrelated. I'm here to tell you otherwise, that they all stem from flaws in our information system. Information is the lifeblood of any society, and our three-decade-old digital system for distributing it is fatally corrupt at its heart. It has failed to function as a trusted, neutral exchange of facts and ideas and has therefore catastrophically hindered our ability to gather respectfully, to debate, to compromise, and to hash out solutions to the many problems we face, big and small. Everything, ultimately, comes down to our ability to communicate openly and trustfully with one another. We have lost that ability—thanks to how the internet has evolved away from its open, decentralized ideals.

To be clear, when we refer to what the "internet" is doing to society, we are talking about the lived experience of the internet, the way we interact with that giant space of information and content that we tap into every day, *not* the physical infrastructure that connects the devices through which that information flows. That infrastructure remains more or less within the decentralized, no-one-in-charge structure that was built out in the 1970s by a group of well-intended computer scientists from a few U.S. universities. The internet's problems exist higher "up the stack" at what computer scientists refer to as the "application layer" with what we now recognize as the "platforms" controlled by the Big Tech companies of Silicon Valley. Their dominance arose out of the changes that came in the 1990s with the arrival of the World Wide Web. (In this book, we will refer to this period interchangeably as the "web-era," "current," or "second-generation" internet.)

In this phase, the one we're still living in, "the internet" became centralized, controlled by these corporations. It fueled the failure of our information system.

This failure has resulted in social regression. We feel a loss of control over our lives because we are now subject to what I call "digital feudalism"; it's as if we were marching back toward the Middle Ages. Our sense of agency has been seized from us by a clutch of corporate behemoths. You know their names: Alphabet's Google, Meta's Facebook, the Chinese company ByteDance's TikTok, et al. In the pages ahead, I will make it incontrovertibly clear that those companies have done great harm to us. We'll discover how, by scraping and stealing our data, hiding the influence of their proprietary algorithms, and persistently surveilling our online activity, they have stripped us of our agency, of our capacity to set the course of our lives. I'll also show that they've completely undermined our collective ability to drive progress and that, as bad as all this sounds, if we don't move quickly to fix things, it will get much worse as the age of artificial intelligence (AI) rapidly takes shape.

The situation is urgent. We must change this system. Do it right, though, and we will finally, properly unlock the immense potential of the internet, the innate possibility that comes with interconnecting everyone on the planet. We can chart a path to a new, human-friendly internet and a better world.

But we must choose wisely.

~

This is a book about the future, but it begins in the past—on January 10, 1776, to be precise. On that date, six months before the drafting of the Declaration of Independence, an anonymous pamphlet with the title *Common Sense* began circulating from a publisher in Philadelphia. Then, like now, society was marked by a deep and growing unease. For more than a decade, colonial Americans had been chafing at how they were being governed by the English monarchy. Their discontent had climaxed with the Boston Tea Party rebellion in 1773, when the Sons of Liberty rose up against the Tea Act passed earlier that year to demand the right to self-rule with the protest cry of "no taxation without representation."

But while anger and dissatisfaction were widespread, no clear, unified movement had yet emerged dedicated to the radical idea of breaking with Britain and setting up a new government based on democracy and self-determination. Even if they despised King George, many colonists could not imagine living in an independent, people-powered republic. After all, monarchy was the only model of government they had known. Many could not conceive of the kinds of civil rights they would claim under a democracy.

It was a historic inflection point. Writers and storytellers are instrumental at moments like this. As the historian Yuval Noah Harari reminds us, stories are the stuff of civilization. They help people imagine the previously unimaginable and create new forms of collective organization. That was what the writer of *Common Sense*—revealed three months later to be the activist and pamphleteer Thomas Paine—achieved. In its plain, to-the-point critique of the British monarchy, *Common Sense* made an irrefutable case for independence. With sharp language that called out lies and ruthlessly exposed the inherent flaws of an autocratic system, Paine brought into view what had been hidden in plain sight. From then on, his readers could no longer unsee it:

Paine showed that British rule was based on an oppressive and exploitative system. And with that, many abandoned their loyalty to the British Crown.

In a mere forty-seven pages, Paine broke down the muddled logic of a monarchy's divine right to govern and offered an alternative vision of democracy and self-rule. He was not overly prescriptive on the precise form of a future government, since that should itself arise from the will of the people. But he was emphatic that the first step must be to break free of the Crown's corrupt and self-serving constraints.

At the time, many American colonists were looking for compromise, sending envoys on the long voyage to London to petition King George into softening his position. The reality, Paine noted, was that "our prayers have been rejected with disdain." He was now convinced that "nothing flatters vanity, or confirms obstinacy in Kings more than repeated petitioning." Following the brutality of the Tea Party's suppression, reconciliation was "a fallacious dream," he wrote, leaving separation as the only viable path for Americans to build the bright future for which they were destined.

~

Common Sense had a massive, world-changing impact. Various accounts put its sales at more than 100,000 copies in its first three months and as high as 500,000 by the end of 1776, accounting for one-fifth of the colonial American population. Landing at a pivotal moment, it played a powerful, catalytic role in driving the movement toward American independence.

Paine's pamphlet was published during a time with distinct

echoes in the present day. Now, as then, we face the unavoidable reality of a broken system, a sight we can no longer unsee. Taking inspiration from Paine's approach, this book will lay out the irrefutable truth of that system's failure.

Some of this might be difficult to hear. I will argue that as much as we like to think we are in control of our online choices, we often aren't. We have allowed the major platforms to rob us of our data and to subject everyone to an insidious system of dangerous behavior modification. Still, my outlook remains hopeful. Because in recognizing the need to overthrow that system, we can also embrace a solution.

I'm far from alone in highlighting the harm the current, web-era internet has done to twenty-first-century society. This conversation has been going on for a number of years. Proposals to fix the problem run from aggressively regulating Big Tech to replacing platform-centralized Web 2.0 social media with decentralized Web 3.0 versions giving users greater control over what they read and watch. These are all commendable efforts to reform a broken system, but they don't get to the core of the problem. Like Paine, I'm not calling for a patchwork solution. We need a fundamental change to how the internet handles a highly valuable asset that emerged out of the web era: the data that's captured in our interactions with one another. We must overhaul the system of control over that data—data that is rightfully, morally ours—and take charge of it ourselves to reduce our dependency on these feudalist platforms. In so doing, we can diminish their and others' ability to use that data to manipulate us, harm our children, and undermine our democracy. And we can deplete a key source of their power and unprecedented wealth. Central to my thinking is that we must flip this power dynamic and give every one of us our rightful claim to participate in the value, whether monetary or otherwise, that the internet's connectivity should enable in the world. This is not some dry, dispassionate issue. In the web era, I will argue, our data is intertwined with our human essence, our very personhood. Gaining control over it is a matter of asserting our human rights.

There can be no more petitioning the king, which finds its modern-day equivalent in congressional representatives seeking guidance from Big Tech executives on how best to regulate their industry. If the feudal lords get to dictate superficial policy responses to the abuses we've suffered—whether it's the loss of privacy, the manipulation of voters, or the harm done to young people by social media toxicity—the problem will not be fixed. And as we enter the age of artificial intelligence, the major platforms' control over our data and, by extension, over our lives will become absolute. Like absolute monarchy.

This book will take as its guide the form of government and the resulting social structure that Paine and his fellow founding fathers dreamed into existence—a set of ideas I will call the American Project. I believe that, for all its faults, the U.S. model of society and government provides a unique set of principles for assessing our current tech system's failings and for designing a superior replacement. The United States has always been an aspirational work in progress. It was imperfect from the outset, and still isn't perfect. But its founding principles provided a framework for its citizens to resolve differences and, ultimately, build prosperity and advance the cause of freedom. Sadly, we have lost our way. I will make the case in this book that our democratic foundations have been undermined by autocratic, centralized surveillance technology. It's up to all of us to fix this and get the American Project back on track.

Unlike the revolutionaries Paine stirred into action, we need not shed blood this time. And the systemic change we require doesn't depend on action by politicians, too many of whom have been co-opted by Big Tech's money. Since the core problem is the web-era internet's technological design, the solution must be partly technological. But that's the easy part, relatively speaking. The ultimate agents of the change won't be the coders or software engineers, many of whom are already deeply engaged in the repair work. As I write this, they're developing new protocols, building new interface tools, rolling out new data storage systems, and fostering a new model of digital identity that gives users authority over their personal information. No, we the people must make this change happen. We can do this by choosing to support those projects and then to *migrate* to a new, upgraded version of the internet designed to serve the interests of people, not those of algorithmic machines.

~

"Migrate." There's another American Project word. Hundreds of millions of people from all over the world chose to physically migrate during the two and a half centuries that followed the conclusion of the Revolutionary War and the installation of a new, democratic constitution in what would become the United States of America. They were drawn by the vision embedded in the American Project, the idea that in America a citizen had "unalienable rights" (per the Declaration of Independence), that those rights were protected, and that individuals were respected. As American citizens, they could take charge of their life and strive to make something of it.

This vision has meaning for me and the other author of this book. Both Americans, we share similar roots but had quite different journeys to citizenship.

I descend from immigrants who fled Ireland's potato famine in the 1850s and settled in the Boston area. In a story that encapsulates the American Dream, my ancestors built on those humble beginnings to forge a successful, multigenerational construction business. From that base, I had the opportunity to expand the company's remit, launch my own businesses, and help develop and enhance some of the foundational infrastructure on which America's cities depend.

Through business and civic entrepreneurship, I've sought to leave a positive impact on society. Recently, I've focused these efforts on an organization I founded called Project Liberty, a broad-based initiative that is mobilizing people and technology to create an open, human-forward internet infrastructure that positively impacts our daily lives. Put simply: Project Liberty is building a better web and, in turn, a better world for all of us.

Some might wonder what drove a guy who started his career working with bricks, mortar, and cement to dedicate so much of his time to reengineering how the internet works. After 130 years of my family constructing solid, safe infrastructure that's designed and built for human beings—including laying a good deal of the fiber-optic cable networks on which the internet now runs—I've come to chafe at Silicon Valley's "move fast and break things" ethos and its disregard for the people who use what it makes.

I'm also the father of seven children, which inspires in me a strong desire to put the internet's infrastructure on a more solid foundation for the future. The four from my first marriage lived through a difficult period at the end of my ownership of the Los Angeles Dodgers. At that moment, the media circus around my messy divorce—a mess for which I share much responsibility—revealed to me how dangerously powerful social media can be as a tool for misinformation, manipulation, and character assassination. I acknowledge, and am saddened by, the pain that my four sons endured and am motivated to help fix the toxic system that exacerbated it. I also want to ensure that my younger three children, whom I'm raising with my wife, Monica, never experience anything remotely similar. But this, of course, isn't just about my children. This is about all children, their future, and the world we will leave for them.

To me, this is also about honoring my ancestors, who created something lasting and real for their descendants and many others. As builders, they quite literally helped make the American Project a reality. And I recognize that all of you reading this book also have ancestors worth honoring.

Around the same time my ancestors took the trip to Boston from Ireland, those of my coauthor, Michael Casey, fled the same famine in Ireland for Victoria, Australia. After traveling widely and working on five different continents, this Australian decided to become a naturalized U.S. citizen in 2003, not because he had to—after seven years of marriage to his New York–born wife, Alicia, he had a perfectly valid permanent green card—but because he wanted to participate in American life and share the experience of citizenship with his wife and daughters.

Michael's passion for this issue is shaped by that outsider-insider experience but also by his long career as a journalist, which began before the internet and has spanned its evolution.

Throughout, he watched with dismay as his and his colleagues' efforts to attract audiences to important, well-reported stories were hijacked by the clickbait-favoring algorithms of social media. His dedication to quality journalism and his deep interest in the role that technologies can play in elevating the truth and bolstering societal trust make him the ideal collaborator in this project.

~

One fact we wish to address: This is a book written by two well-to-do, over-fifty-five white men. Just as all books are shaped by the lived experiences of their authors, we acknowledge that this one, whether we like it or not, comes from a position of privilege and with all our inherent biases. Still, we expect that readers will find the situation we are describing entirely familiar and that, while all of us have a different experience with life in the internet age, the solutions we're proposing are relevant and available to everyone. The essential point here is that we need to devise a model that both elevates and protects the rights and responsibilities of all human beings, regardless of their background, and works to strengthen the society to which we all belong.

Let's also get out in front of one of the straw-man arguments that Big Tech might throw at us: Michael and I are not Luddites. We are not anti-tech. We want to unlock all the incredible problem-solving capacity that new technologies will bring to us. And we're certainly not opposed to information technology, the most important tool a society can develop. We simply want that technology to be built and operated in the interests of human beings. And the problem, as we see it, is that many internet-based technologies that arose at the turn of the millennium made humans subordinate to machines.

Separately, we'd like to address a more sympathetic group of technologists, including some who've generously contributed vital work to Project Liberty and who may wish we had gone into more technical detail on certain topics. We felt that a book addressing the most important issue of our time needed to be written in a style accessible to the widest possible audience. That meant that, wherever possible, it should avoid technical terms from the worlds of web development, cryptography, and blockchains that could prompt a lay reader to tune out.

Some computer scientists might find our descriptions simplistic. We know full well that the internet's founders thought about human users when they created it; our point in stating that the internet currently favors "machines over humans" is that it was allowed to develop in a way that put algorithms and devices in charge of our lives and that we must now optimize the internet for people. Also, while we may from time to time talk about a "new internet," we know we can't create an entirely new architecture but rather must evolve the current one to a next generation that serves society. Additionally, we acknowledge legal, technical, and other challenges to the idea of literally "owning" one's data but felt such language could help to spur people, with skin in the game, to take charge of their lives and fight for their rights. We're grateful for the technologists who understand the need to reach broad audiences, to encourage a popular movement that can adopt the projects they've worked on so that they can meaningfully change the world. In fact, we're relying on those computer scientists who agree there's a problem with the current technology to help us fix it.

Finally, while this is a book written by two Americans, with frequent references to the American Project and to the current U.S. economic and political systems, its implications are global. The internet is borderless by design. The opportunities and problems it generates apply to everyone, and the aspirations of all who use it are not bounded by nationality or ethnicity. The values enshrined in the Declaration of Independence and the Constitution, including the Bill of Rights, were conceived as universal principles and were built on the liberal Enlightenment ideas developed by English, German, and French thinkers. They've also underpinned democracy movements from the Middle East to East Africa to Southeast Asia to Latin America and beyond. The same universality applies to the human rights we must now assert in the digital realm. If we're going to overturn the old internet system by migrating to a new one, that new version must respect all of humanity.

Having said that, Americans constitute an especially important audience. The internet was a U.S. invention. And most of the companies that dominate it are American, with the exception of a few in China. We U.S. citizens and our institutions have an especially big responsibility to right the current system's wrongs.

Throughout the previous century, the United States led the world economically, politically, and technologically. Now, as we build a new, open internet that works for humanity rather than against it, Americans have an opportunity—perhaps the last one we'll ever get—to lead the world out of a mess we helped create.

In the pages that follow, we will show how righting this wrong will require wresting control over our data away from a small group of Silicon Valley–funded companies. Digital data can seem like an obscure, bland, and rather unsexy issue, but consider what's really at stake here. It's about who gets to define who you are. Every time we mention data in this book, what we're really talking about is *you,* your personhood in the internet age. Think of your data as your digital DNA, your lived experience in this digital world we now occupy. It is as important as the biological DNA you're born with, if not more so. The issue, then, is that in controlling your data, the Big Tech companies have stolen your life from you. As of now, you don't own you. *They* do.

If we approach this mission with the spirit of the American revolutionaries in 1776, we find strong parallels between the values on which they founded the American Project and those that must underpin a better world for all of us in the internet age. In this book, we address those values within a framework we call the Four Rs: *rights, responsibilities, rewards,* and *rules.* Each item in this framework corresponds to a separate set of features defining our sense of self, our relationship with one another, our place in the economy, and our choices about how we are governed. Thus, in separate chapters, we address our *rights* as individuals, our social *responsibilities* as citizens, the *rewards* or incentives that power the market economy in which we strive, and the *rules,* laws, and protocols to which we choose to be subjected. We must fundamentally rethink each of these for the internet age.

Before we get to the Four Rs, though, we need to go to the heart of the problem by examining the flaws in the current internet, the harms it is doing to individuals, and the damage it has imposed on society. That's the focus of chapter 1.

By now you'll recognize that we're not dealing with a small matter. The central idea of this book transcends the nitty-gritty details that software engineers grapple with when amending computer code. It's about reforming society in the digital age. It's about relearning how we treat one another so that our collective interests are advanced. It's about saving humanity.

CHAPTER ONE

The Heart of the Problem: An Obsolete Internet Design

A war in Eastern Europe. A war in the Middle East. The hottest summer in recorded history. An opioid epidemic killing three hundred people a day. Tens of thousands of migrants teeming across the southern border. Train derailments. Mass shootings. Pothole-riddled highways. Surging interest rates pushing the dream of homeownership out of reach for many.

This snapshot of American life, from late 2023, shows a house that's well and truly on fire. The response in Washington, you'd be forgiven for assuming, would be a desperate, unified effort to fight the flames. Yet the conflagration coincided with a leadership vacuum: A rabble-rousing segment of the fractured majority party in the House of Representatives had just ousted its Speaker, leaving the urgent business of government at a standstill for weeks.

Amid it all, a profound mistrust of politicians and the institutions of government was growing ever more entrenched. As the country approached the 2024 presidential election, polls showed that most Americans had unfavorable opinions of the two front-runner candidates—one a twice-impeached former president facing four criminal indictments, the other a sitting president confronting his own impeachment threat in the House. Regardless of who the nominees would end up being, everyone was bracing for a tense, yearlong slanging match consumed with hyperbole, mockery, ad hominem attacks, and an utter lack of nuance, compromise, or middle ground—all with the threat of violence hanging ominously in the air.

With these extreme, often brutal divisions penetrating all layers of society, from state legislatures to town councils to school boards, several nagging questions loomed: Why do we seem so ungovernable? Why is our decision-making apparatus paralyzed? How did we become singularly focused on one-upmanship and gotcha tactics rather than on the common need for effective, consensus-based policymaking?

The answer lies in the failings of our information system. A healthy democracy depends on a healthy system for producing and distributing information, and the truth is that the one we've come to rely on over the past three decades—the sprawling global network of interconnected computers we call the internet—has become decidedly unhealthy. As currently structured, the internet enables the secretive, centralized aggregation of massive troves of personal and social data generated by our online interactions. That system has empowered a small group of tech companies and their clients to exert untold influence over us. With hidden, proprietary, and self-updating algorithms that are constantly learning from our data, they curate the torrent of content flowing through social media that has become the primary source of information for billions of people. In doing so, they've learned how to tap into our most basic instincts to engender the conditions that maximize our engagement with their online platforms and, by extension, their profits. Some time back, they learned that engagement is most easily optimized when we are triggered. So, inevitably, their algorithms—a form of artificial intelligence—built a machine that would encourage the kind of toxic, antisocial behavior that's destroying the fabric of society.

This digital system extends far more deeply into our lives than any of the information systems on which earlier civilizations depended, far more even than the internet's early designers expected when they started linking hardwired mainframe and desktop computers to one another in the 1970s. Now this system encompasses all the connected devices we interact with—from our smartphones to our TVs to our cars to our home security systems—forming a network called the Internet of Things. This network even provides the backbone of our financial system and of the global economy's supply chains.

The internet has become inseparable from life itself—to the point where our DNA, our fundamental biological makeup, is now also fully digitized. In a recent conversation I had with Broad Institute director, Todd Golub, a world leader in applying genomic tools to cancer research, he explained that the cost of the first sequencing of the human genome in 2003 was $3 billion. Now, with the cost approaching $100 per person, every child born in the developed world will soon have their genome sequenced. Todd asked me to consider three questions: Who will own that individual's data? Where will it be stored? And who gets to use it for what purpose? These questions are fundamental because, as we noted in the introduction, there is no longer any separation between our digital and our real-world existences. Again, whenever we say "your data," think "your personhood."

Where does this leave us? We now find ourselves in the battle of our lives, a battle over the capacity to *own* who we are, to define our identity, to live as we choose, and to do so in a way that fosters a healthy coexistence with others.

This is not some wacky conspiracy theory about a cabal of "deep state" puppet masters manipulating us behind the scenes. Nor is it Hollywood-imagined science fiction, an imperceptibly fake reality created for us by a band of vengeful androids—though that metaphor is quite useful here. Our predicament is the result of a flawed structure. Through our actions and lack of actions, we have permitted the development of a system that incentivizes corporate behavior and contradicts the broader interests of society. We can, and we must, change that model. This starts by recognizing and addressing the mistakes we made, the first of which was to grant a few monopolistic internet platforms access to an inordinate amount of personal data about us—*our* data, data that must be put under *our* control. The second was to let those platforms use that data to determine what information we see, hear, and absorb, and to allow them to do so in pursuit of *their* interests, not ours.

Data is the lifeblood of the digital economy, more valuable than oil. Given the value of data, today's tech giants enjoy an outrageous, unfair imbalance of power in their favor. They have the capacity to shape every aspect of your life: your job, your kids' education and well-being, your healthcare. They've converted that power into the most impressive wealth-accumulation machine that Wall Street has ever seen. Throughout the three decades of the internet in its current, platform-dominated manifestation, the five biggest internet companies have grown to a combined market valuation that's larger than the gross domestic product of any country other than the United States or China. For the rest of us, that same period has produced, on average, slightly higher real incomes—but at the cost of being overworked, overstimulated, and overstressed. Since the World Happiness Report's survey of different countries' emotional well-being began in 2015, the ranking for the United States has been sliding ever lower. It's not all in your head.

It doesn't have to be this way.

We can fix this. In fact, we must. We have arrived at an inflection point. Every day more people realize that our house is indeed on fire. Perhaps it's the crisis we need, the kind that creates opportunity. It gives us a reason to go deeper than the surface-level problems to find their root, their common cause, to drill down into the core system that underpins all of it. But if this is a moment of opportunity, it is also a moment of great urgency, for it is about to become even more complicated by the explosive expansion of generative AI, a technology that Goldman Sachs estimates will replace 300 million jobs. The difference between generative AI and traditional AI is that the former can create something new rather than just perform preset tasks—think creating original art versus playing chess. The power of this new breed of AI, thrust into mainstream attention by OpenAI's ChatGPT system, will eclipse that of the AI algorithms that have driven our social media feeds. We can use all this power for good, or we can succumb to it. Our humanity and, indeed, our very lives are at stake.

An Unhealthy Information System

Society is shaped by information. Throughout history, big shifts in the organization, delivery, and consumption of information have had sweeping impacts on the shared experience of life on this planet. The invention of the internet was one of the most dramatic of these moments, but it was far from the only time when a new information technology changed the course of history. Spoken language, cave paintings, and writing were cognitive and technological breakthroughs that moved humankind out of a primitive existence into what we now call civilization.

As cities, kingdoms, and states formed, control of information became a source of great power. Historians study past civilizations to find correlations between control over the information system and the degree to which power was distributed. Was access to information shared evenly? Was technology used to encourage the positive, open exchange of ideas? Or was information concentrated narrowly in the hands of a few, who, in order to entrench their position and expand their power, used technology to encourage narrow, closed-minded ideas? Did the information-sharing technologies support the common interests of the many or those of a few elites?

The history of one country in particular—Germany—offers valuable lessons in how information technologies can allow for an even distribution of power or concentrate it in the hands of a few.

Mainz is a small city on the Rhine, not far from Frankfurt. There, Johannes Gutenberg, inventor of the movable-type printing press, was born, somewhere between the years of 1393 and 1406. In his youth, books would have been a rare thing in Mainz or, for that matter, in any town anywhere in the world. Most villages in Europe likely had only one: a handwritten Bible, probably in Latin or Greek, held in the possession of the local priest, the only person among the townsfolk who could read. When that robed man read to his congregation from the holy book, it felt like he was a conduit of divine information, the envoy of the Almighty. Most listeners would not have understood the Latin phrases the priest was intoning, but that would only have added to his mystique and power. For centuries, this was the centralized, top-down, dogmatic information system of the Middle Ages.

Gutenberg's printing press changed everything. Introduced in the mid-fifteenth century, it set in motion a gigantic shift in the power dynamic. The technology enabled widespread production of translated Bibles in multiple languages. Their increasing ubiquity encouraged a greater interest in them, which led enlightened scholars to embark on efforts to teach people to read, setting European societies on a path toward mass public education in the centuries that followed. That, in turn, encouraged critical thinking, including the thought processes that led Martin Luther to devise his "Theses" challenge to the Catholic Church, which he printed, published, and replicated with the help of Gutenberg's press to give rise to protestantism. As literacy expanded, demand for more reading material grew, which created an incentive for the production of non-biblical texts, spreading new ideas and new ways of seeing the world. All of this whittled away at the domination of the church and the priest. It paved the way for the Renaissance, the Enlightenment, the Industrial Revolution, and the modern era. There are few technologies that can claim such a powerful, positive impact on humanity.

Now let's go to Rheydt, 140 miles north of Mainz, not far from Düsseldorf. There, in 1897, Joseph Goebbels was born. After Adolf Hitler's ascent to power, Goebbels became the Third Reich's minister of propaganda. Goebbels was the architect of the most evil violence- and hate-inciting information system the world has ever known. Mastering the relatively new technologies of radio, film, photography, and graphic printing, he and his staff produced the words and images that bred the Aryan supremacist and anti-Semitic ideas that would justify the unjustifiable. In 1945, upon the collapse of the regime at the end of the Second World War, in which an estimated seventy million people perished, including six million Jews who were targeted through genocide, the Nazi official arranged to have his six children killed with cyanide before committing suicide with his wife, Magda.

The stories of Gutenberg and Goebbels underscore that information technologies are tools. Just as a hammer can be used to build a home or to kill someone, they can be used for wonderfully constructive, life-affirming purposes or unthinkably evil ends. It all depends on who is controlling them, which in turn depends on the power structure built around them. That distinction is even more critical when we're talking about technology that gives rise to an entirely new information *system*, which is what the internet delivered. The foundational design principles of such a system matter enormously in how those power structures form over time.

The internet has massive potential to do good. It can hammer a lot of nails. Given its undeniable utility—the fact that we can use it to navigate while driving, remotely manage our homes' thermostats and alarms, easily connect with loved ones far away, and instantly access news, music, and videos in any language from anywhere in the world—the idea of returning to a pre-internet existence is just not an option. But the internet's current structure enables an excessive concentration of power in the hands of a few dominant tech companies, and they are doing more harm than good. We must fix that. We need new technologies that give the people who use them much more power to determine what is read, heard, and viewed over the internet, all while maintaining the interconnectivity and ease of information-sharing that give it its value. We need a new design, one that converts an autocratic, antisocial surveillance technology into a democratic, prosocial technology that respects our rights as humans.

If we can do this, if we can put data into the hands of the people to whom it rightly belongs, we can start to tap the incredible innovations promised by digital technologies without fear of harmful side effects for society. Imagine wearing a device that tracks your health vitals 24/7 and safely shares privacy-protected data with an AI healthcare adviser in real time. Imagine an electricity grid so rich with reliable data on power usage and solar generation that it can efficiently provide clean energy to every home at near-zero cost all day long. Imagine an abundant knowledge pool that scientists anywhere could tap to develop food in a lab and end starvation or to globally share 3D-printed blueprints that could help house the world's homeless. Imagine all of this, without the harms.

In accelerating solutions to the planet's many urgent problems, we believe new technologies like generative AI and quantum computing, if deployed safely, can usher in an era of unprecedented prosperity and well-being. On the flip side, if we don't act, the societal dysfunction described at the outset of this chapter will only get worse, especially as AI starts to make disinformation even harder to control, gives bad actors more power, and expands the opportunities for tech companies to abuse our data and content. Without a fix soon, we will continue our rapid descent into authoritarianism and social dysfunction. Again, it's our choice.

Lessons from the American Project

So how *do* we fix the internet? Well, not with Band-Aid solutions. Previous approaches to solving the internet's ills—for example, the European Union's privacy protections, such as 2016's General Data Protection Regulation (GDPR) and its "right to be forgotten" laws—have not delivered on their promise because they failed to address the core problem. As the great architect and design philosopher Buckminster Fuller once said, "You never change things by fighting the existing reality. To change something, build a new model that makes the existing model obsolete." That's the mindset we need to bring to this most urgent of tasks.

A foundational overhaul starts with recognizing a basic challenge that has long dogged the engineers who've built and maintained the internet: how to safely allow people to privately identify themselves to one another without exposing sensitive information to the world at large. The internet's open-information architecture is what provides its value to humanity, but it also creates potential for mass surveillance. That identification challenge was much less of a problem for the nonhuman machines that the internet's founders hooked up to the network, the computers charged with coordinating its decentralized system for sharing "packets" of data. Using something called an internet protocol (IP) address, each was assigned a unique identifier whenever it connected to the network—and that system functions the same to this day. Whether it's your laptop, your smartphone, or your TV connected to Wi-Fi, it's the device that's identified with that all-important IP address, not you. This wasn't much of an issue during the first, relatively sleepy two decades of the internet's life. But in the 1990s, when giant waves of nontechnical ordinary folk hooked up to the web and started exchanging money and things of value with businesses and one another, we needed to know we could trust the humans on the internet, not just the machines. Who was going to oversee the verification of these people? And how? Let's just say the way this challenge was resolved was less than perfect. It left all of us, the human users of the system, beholden to powerful intermediary institutions that vacuumed up our data within an opaque, closed system. The intermediaries that ultimately became dominant were the giant internet platforms that control our lives today. It's now up to us to come up with a system that wrests back control.

In the chapters ahead, we'll lay out some proposals for a new protocol model that re-empowers humans, ideas under development within the Project Liberty initiative. These are proposals, not dictates. It's up to the market to decide which software becomes standard and which new innovations are adopted. Nonetheless, the principles embedded in our ideas should help us find solutions that serve the common good. They offer a solid foundation on which to imagine a better future.

Whatever structural redesign wins out, implementing it—and replacing today's irreparably broken structure—is an urgent, unavoidable first step in fixing our fractured society and our polluted civic discourse. Merely introducing new laws and regulations won't cut it; they will never keep up with the power and pace of new technologies. We must upgrade the plumbing of this digital information system, augmenting its application-layer protocols with new software to overhaul the data-hogging system that goes with it. Whatever we come up with must have humans front of mind.

Still, while both the problem and the solution lie with

technology, the catalyst for change must come from everyone, not just from the engineers and developers. We need popular participation, a collective mission that transcends the narrow, functional questions of software design. An overhaul of the internet is inherently a social and political project.

To lay out a vision for this brand-new project, we'll be viewing the problem through a lens afforded us by a much older one: the American Project, set forth by Thomas Paine and his fellow founding fathers. While there is no perfect society, we believe that the American Project—at least as it was designed, implemented, and iterated upon for most of its two and a half centuries of existence—offers a useful benchmark against which to conduct this analysis. For all of the United States' flaws, including an era of legalized slavery, no other society has produced the same level of prosperity, technological advance, intellectual and cultural dynamism, and freedom in such a short period of time.

We recognize and respect that billions of people worldwide choose to live under different systems. But the core principles of the American Project were based on ideas that were conceived as universal. In fact, the Universal Declaration of Human Rights, adopted by the United Nations in 1948, includes many of the same principles. We have found the concept of basic human rights and fundamental freedoms to be a useful framing device for comprehending and addressing the problems we face in the internet age.

The American Project can be viewed as an amalgam of rules, social norms, and institutions intended to allow a nation to grow, address changing circumstances, and perpetually generate opportunities for advancement among its citizenry as well as to ensure a general state of peace and well-being for everyone. With democratic underpinnings founded on principles of personal liberty, equality under the law, and free markets, it cultivated an understanding among American citizens that while they were privileged to enjoy those core personal rights and freedoms, they had to respect the same for others. The American Project fostered a parallel responsibility to look out for the common good. Most important, it wasn't conceived of as a finished concept; it was always a work in progress. It was built to enable change, with U.S. laws expected to evolve with changing circumstances and shifting mores, while government leaders, legislators, and the courts treated those founding principles as the North Star against which present conditions should be assessed. It's how the nation's electoral rules evolved from "one white man, one vote" to "one person, one vote," and why our society continues to confront, debate, and address areas in which it is thought to fall short of those core ideals. Without this framework, progress would be impossible.

Human Data Mines

Before the Enlightenment spread across Europe, inspiring the likes of Thomas Paine and Thomas Jefferson in colonial America, most people were *subjects.* Their claim on life was quite literally subject to the discretion of a king or queen, and their livelihoods were determined by an accident of birth that left them obligated to serve the lord or baron on whose land they toiled. Under American democracy, people became *citizens,* a concept that not only recognized their rights to liberty and the pursuit of happiness but also ensured that their votes—not some specious notion of divine right—would determine who governed them. It was a direct manifestation of the philosopher Jean-Jacques Rousseau's notion of popular sovereignty, which proposed that governments have a legitimate claim on power only when they are derived from the "general will of the people," a situation that creates what he called a "social contract" between the government and the people. In this way, the rights and responsibilities conferred on the new citizens of the United States of America would collectively empower them to determine a shared destiny for their nation. Citizens would have agency—a capacity to make choices and effect changes—whereas subjects did not.

In contrast to these core principles of liberty, our current reality is, in our mind, best described as digital feudalism. Like poor, powerless subjects of monarchs and aristocrats, we are serfs, subjugated by a small group of companies that have exploited a feudal internet architecture. In this system, human beings are treated as afterthoughts—or not thought of at all—in service of building massive data extraction platforms. In the United States, the ruling clique comprises our era's biggest software companies: Google's owner, Alphabet; Facebook and Instagram's owner, Meta; Amazon; Apple; and Microsoft. The latter, with its giant investment in the research organization OpenAI, has joined the tech arms race to control our data, and the power and profits that come with it. The founders, senior executives, and large investors in the Big Tech companies command great sway over the operations of the internet. Thanks to them and their management teams, whose compensation models incentivize them to maintain or double down on the status quo, we live at the discretion of their proprietary algorithms. The software programs based on these algorithms—which, much like computers, servers, and devices, should be understood as machines—treat us as quarries from which to mine *our* data, now the most valuable commodity of the digital economy. They then aggregate and organize this data and use it to create tools with which their corporate leaders can influence us.

Aided by sensors and data-capture points positioned at every turn in our daily life—in the words and emojis we put into texts and social media posts, in our purchases, in the movements of our computer mice, in the cameras pointed at us, in the GPS devices that track us, in the listening devices in our homes, and in the music to which we listen, the videos we watch, and the photos we share—these machines store far more information about us and our social connections than our own brains could possibly store. As one MIT professor put it to me, "We are living in a minimum-security prison. We just don't know it."

According to a 2016 ProPublica report, Facebook at that time collected an average of fifty-two thousand data points on each of its users. That number, now likely very much higher, gives a sense of how we are viewed by the platforms. Their "black box" systems, built with code into which no outsiders have visibility, extract a valuable commodity (our data). They then use that commodity to assign us each a profile and to fabricate a powerful machine (a proprietary algorithm) with which they categorize, target, and manipulate us. In doing so, they systematically dehumanize us—much as the feudal system dehumanized peasants.

In all of this, there is no social contract or even a moral obligation for these platforms not to treat us as their pawns.

Instead, they've buried us in legal contracts, most of them in fine print that no one reads, imposing terms and conditions surrounding our use of these applications that compel us to forgo any claims over our data and the content we create and post. We've literally signed away our rights and surrendered our personhood to these Silicon Valley giants.

This dynamic has been building for two decades, but only very recently have more and more people started to recognize the enormity of what we've given up. A seminal work in this field, Harvard professor Shoshana Zuboff's 2018 book, *The Age of Surveillance Capitalism,* recounts how the data extraction model originated at Google in the early 2000s. Facebook, after discovering it could profit from a powerful self-reinforcing feedback loop, adopted and updated the model. Essentially, this is how it worked: Facebook's surveillance of its users' activity generated insights into how people responded to different textual, visual, or aural stimuli. Facebook's data scientists and engineers then tweaked the platform's content curation algorithm in a bid to steer users into engaging with other users for longer periods of time. Internally, Facebook called this engagement meaningful social interactions (MSI), and the MSI metrics served Facebook's and its clients' revenue goals. The cycle then repeated over and over, as new behaviors generated new data, allowing for iterative, perpetual "improvement" (i.e., more precise targeting) in the algorithm's ability to modify user behavior.

One of the most notorious applications of this arose in the Facebook–Cambridge Analytica scandal, news of which broke in 2018. For years, the British consulting firm Cambridge Analytica collected Facebook users' data without their consent and used it to feed them microtargeted disinformation. One of the goals was to influence the 2016 U.S. presidential election; another was to sway the United Kingdom's Brexit vote. But, in many ways, that high-profile scandal was an outlier: A far bigger, if more subtle, problem is the data extraction model's impact on our day-to-day lives. In multitudinous ways, these platforms' algorithms color our view of the world, shape how we react to issues that matter, and drive us into the hands of advertisers. Zuboff says this exploitative business model, which has migrated from Facebook to become the modus operandi of virtually every internet platform or application, has stripped us of what makes us human: our free will, without which neither democracy nor markets can function.

Perhaps you're sitting there thinking, "Nah, this isn't me. I'm in control. I can't be swayed by some computer code. I'm open to all ideas and suggestions, and I deliberate on them, carefully weighing the pros and cons of each before deciding what to do."

We hear you. There are various areas of our day-to-day lives over which we retain control. But they are dwindling, because powerful interests profit from depriving us of that control. The owners of these tracking and advertising-maximization systems have not spent the past two decades figuring out what makes us tick for nothing. They've watched to see what content suggestions provoke the dopamine releases that lead us to click, to "like," to follow, or to share. They've figured out our political leanings, our artistic tastes, our sleep habits, our moods, and, most important, the social groups and online tribes with whom we form connections and allegiances. Facebook, it is said, knows you are going to break up with your partner before you do. If you even briefly let go of the "I'm in charge; no one is telling me what to do" mindset, you can see how the platforms can and will use their gigantic data hauls to shape our individual thoughts as well as our collective behavior—because they are incentivized to do so.

Here's another way to think about how you pay for all this data extraction: a two-decade divergence in prices for different goods and services in the U.S. economy. A chart from the online publisher Visual Capitalist shows how prices for goods and services that you need to live a healthy and productive life—such as medical care, college tuition, housing, and food and beverages—all rose very sharply between 2000 and 2022. By contrast, prices for products that integrate with the internet and extract our data—such as software, cell phone services, TVs, and other entertainment devices—all fell significantly during the same period. It's worth asking why this is the case. Your quality of life in the nondigital world is deteriorating, but your digital existence keeps getting oddly less expensive. The reason is that the latter is subsidized by the ever-larger amounts of data you hand over to tech companies.

We need to think harder about the real price we are paying for data extraction devices and related software. Remember, your data is you. Here, the timeless words of the information security guru Bruce Schneier are helpful. Nearly a decade ago, he wrote: "If something is free, you're not the customer; you're the product."

~

To repeat what we said in the introduction, you don't own you; the platforms own you. Keep that in mind as we dive into the four chapters covering different perspectives on what's needed for a healthy information system, each framed around one of the Four Rs. The first one: our *rights* as individuals. It's time to stand up and assert our right to own our data, our right to own our lives, and regain our personhood. We'll discuss this in more detail in chapter 2.

CHAPTER TWO

Rights: Personhood in the Internet Age

Let's just come out with it: In the internet era, you have been stripped of your rights. You have no agency. You are not a citizen. You are a vassal, a serf, a subject, beholden to the demands of your digital feudal overlords. It is exactly the kind of dehumanized status that the founders of the United States intended to banish to the dustbin of history. Surrendering these hard-fought rights shouldn't be the price we pay for simply using the internet. Don't we want an internet that respects, protects, and enhances our rights?

To create a protected space of freedom inside which Americans could pursue happiness and self-actualization on their own terms, the founders deliberately defined a set of all-important rights that, per the Declaration of Independence, were unalienable—that is, inseparable from individuals. They then enshrined those rights in the highest law of the land and built an independent judiciary that would vigorously protect them. In service of the basic ideals of the Enlightenment, these actions placed the individual human at the center of America's social and political structure. It's why the Bill of Rights—the first ten amendments to the Constitution—is viewed almost as a sacred document, a unique expression of the ideals of the American Project. The invasive platforms of the internet have undermined all of that.

In this chapter, in which we explore the first of our Four Rs, we'll lay bare the many ways in which today's dominant digital platforms use their machines to drive us into a mental dependence that's antithetical to the principles of citizenship. As a member of a defined polity, a citizen is supposed to be afforded certain rights, such as the right to express ideas and opinions, the right to vote, and the right to hold religious and cultural beliefs, and is expected to uphold certain duties such as paying taxes and living within the laws. Citizenship conveys a sense of ownership or participation in how a society is run. That's not what we get as users of the internet. First and foremost, that's because we've lost some fundamental rights.

Lost Agency

Have you ever been the victim of data theft? The disruption to your life is immense. We're talking hours and days of lost time, along with mental strain, uncertainty, and a sense of disorientation. For a time, you will feel completely helpless. The platforms and services to which you are now desperately seeking to restore access will put the onus on *you* to prove you are who you say you are. Oh, the hoops you will have to jump through, just to prove you are entitled to *your* information: your bank account, your website, your playlists, your social media posts, your photos and videos.

This is what a loss of agency feels like, and the individual is rarely to blame for the vulnerability. Hackers go after the biggest troves—the vast, centralized stores amassed by platforms and applications—so that they can access thousands of accounts at a time. You're the one paying the price for the failings of those companies' security systems.

What exacerbates the sense of helplessness in situations like data theft is that the platforms have created dependencies across their services. Without access to one of them, you can't get access to another; you need all of them. Alphabet's Google is an example par excellence. There are 4.3 billion Google users worldwide—more than half of the earth's population. And most people rely on more than one of Alphabet's interconnected services: Google Search, Gmail, Google Calendar, the Chrome browser, YouTube, Google Cloud Platform, Google Home's Wi-Fi system, Google Fiber for broadband access, Waze, Google Maps, and the Android smartphone and its app store, Google Play, just to name a few. Lose your Google login, and your access to all those interrelated services is compromised. It's as if we live in one of those company towns where a big manufacturer that employs half the population and provides everyone with transport, healthcare, and many other local services gets its database of resident IDs hacked, preventing anyone on the list of compromised accounts from riding the bus or attending the town's annual Christmas party. The big difference in our analogy is that this internet-based town's community is global. Four. Point. Three. Billion.

When we talk about scenarios like the one above, we tend to think of how the hacker has victimized the internet user. There's a sense of personal violation knowing that someone else, some anonymous faceless individual, is running around with your personal information, your family photos, your professional records, your financial history, and is, essentially, impersonating you. But you know what? Without any of that illegal activity happening, your identity was stolen long ago. The big platforms stole it. That's the reality we need to acknowledge.

Our loss of agency to platforms like Google is not just an inconvenience. It's theft. It's a violation. And for far too many, it is a source of tragedy and unfathomable heartbreak.

Take the story of Walker Farriel Montgomery, a sixteen-year-old from Starkville, Mississippi, who loved fishing, hunting, and football. On the evening of December 1, 2022, Walker was up late, on his phone, scrolling through Instagram when a pretty girl who seemed to share some of the same contacts appeared in his message feed. She reached out to him, flattering him and enticing him with talk about football. One thing led to another and, when they opened up a video chat, she exposed herself and invited him to do the same. But the minute after he obliged her request that he perform and share a recording of a graphic act, the girl disappeared and a stranger entered the chat. The girl's image, manipulated from footage of a porn star, had been a front for a sextortion scam. The scammer demanded that Walker pay him $1,000 and threatened that if he didn't comply, he would send the recording to all of Walker's contacts.

"We're gonna destroy you if you don't give us the money," the scammer told him. "Everybody's gonna disown you. Your life is over." As Walker pleaded for mercy, his attacker started listing the names of those who would see the video. When he got to the teen's mother's name, it was too much. Walker went to his father's safe, retrieved a handgun, and shot himself.

This grim story was revealed after the FBI, in a forensic investigation, unlocked the teen's phone. The ordeal had lasted several hours and, all the while, Walker's parents, Brian and Courtney Montgomery, had been oblivious to the fact their son was undergoing such psychological torture that he turned to what he thought was the only way out. Despite the unfathomable emotional pain the Montgomery family has had to endure, Brian has endeavored to turn the experience into something constructive for the world—for society, for the rest of us. In an effort to hold platforms like Meta's Instagram to account for their role in extortion tragedies, he joined forces with other mothers and fathers with similarly heart-wrenching stories to become an anti–Big Tech activist.

The parents with whom Brian started working have become the public face of a movement to pass the Kids Online Safety Act (KOSA), which seeks to compel social media platforms to maintain a safe environment for children. Some in the group have lost a child to suicide following bouts of cyberbullying. Others discovered that their children died after participating in internet fads such as the "choking challenge." Still others learned that their teen kids overdosed on drugs provided by chat group syndicates. At the bottom of a letter urging support for KOSA, we find the names of more than seventy-five such parents and those of their deceased children, as well as the ages at which they died:

> Deb Schmill, parent to Becca, 18; Maurine Molak, parent to David, 16; Mary Rodee, parent to Riley, 15; Julianna Arnold, parent to Coco, 17; Jennie and Dave Desario, parents to Mason, 16; Toney and Brandy Roberts, parents to Englyn, age 14; Kristin Bride, parent to Carson, 16; Steve and Joann Bogard, parents to Mason, 15; Erin Popolo,

parent to Emily, 17; Shannon Lee, parent to Ashlyn, 16; Blair F. Aranda, parent to Brantley, 17; Bridgette Norring, parent to Devin, 19; Lori A. Schott, parent to Annalee, 18; Christine McComas, parent to Grace, 15; Jeff Van Lith, parent to Ethan, 13; Rose Bronstein, parent to Nate, 15; Jessica Diacont, parent to Jacob, 15; Kathy McCarthy, parent to Jack, 19; Patti Lujan, parent to Lauren, 18; Laura Lynch, parent to Brillion, 18; Krystal Lebofsky, parent to Deja Star, 12; Avery Ryan Schott, parent to Anna, 18; Tammy Rodriguez, parent to Selena, 11; Maggie Taylor, parent to Emily, 17; Shannon Rickson, parent to Madison; Stephen Cahill, parent to Michael, 20; Hanh Badger, parent to BB, 17; Diana Trujillo, parent to Juan, 16; Mary Popolizio, parent to Alexandra-Victoria, 18; Taj Jensen, parent to Tanner, 20; Christopher Wagner, parent to Christopher; Kathryn Williams, parent to Chantel, 18; Bradley Richardson, parent to Tyler, 19; James Ebert, parent to Jayson, 21; Joseph Ryan Gill, parent to Emma, 16; Erica Adams, parent to Daniel; Dani Clewell, parent to Alexander, 18; Krislynn Wells, parent to Chandler, 18; April Tsosie, parent to Melissa, 16; Michael Ourand, parent to Kevin, 20; Jennifer Mitchell, parent to Ian, 16; Brittney Shealy, parent to Keirstyn, 18; Kristine Mandelke, parent to Jonathan, 22; Jessica Williams, parent to Z.B.; Jodi O'Dell, parent to Justin, 18; Georgia Peterson, parent to Gregg, 19; Shawane Miller, parent to Leon; Shailyn Malone, parent to Taylen, 15; Andrea Silvano, parent to Zachary, 21; Kristina Cahak, parent to Morgan, 15; Fran Humphrey, parent to Sophia, 20; Katherine Klingele, parent to Seqouyah, 16; Christopher A. Dawley, parent to Christopher (CJ), 17; Lindsey Thurman, parent to Manuel "Manny," 17; Monica Ortiz, parent to Fidel, 16; Karen Zients, parent to Ian, 16; Sonja Frezghi, parent to Khalif, 17; Patrick Cly, parent to Vanessa, 18; Tricia and Kinyoun Buford, parents to Braden, 22; Karen Alfonso, parent to a child who died at 22; Annie McGrath, parent to Griffin, 13; Judith Rogg, parent to Erik, 12; Sharon Winkler, parent to Alexander, 17; Michael Kuch, parent to Adriana, 14; John Halligan, parent to Ryan, 13; Todd Minor Sr., parent to Matthew, 12; Georgia Peterson, parent to Gregg, 19; Devon Adame, parent to Kayden, 16; Janet Majewski, parent to Emily, 14; Elizabeth Davis, parent to Cooper, 16; Christina Arlington-Smith, parent to Lalani, 8; Hazel Msuorga Lopez, parent to Paris, 14; Melessa Kocsis, parent to AJ, 19; Desiree Hawkins, parent to Noah, 12; Christina Luna, parent to Josiah, 15; Kristie Reilly, parent to Noah, 13; Karey Kleeman, parent to Aiden, 14.

These accounts of children's deaths just scrape the surface of the suffering that's out there. As cases in which lives were literally extinguished, they represent the most extreme examples in which people have been harmed by an information system that actively drives them into dangerous situations, one that strips people of their rights. But abuses don't need to involve deaths or professional scam artists to inflict enormous harm. People have seen their lives upended by targeted shaming on social media, when influential users turned the victim's comments or images, more often than not taken out of context, into some lesson in righteousness or embarrassment.

One example is a young woman named Ashley VanPevenage, who discovered to her horror she'd become an internet meme after a makeup expert friend who'd helped her cover up acne posted the before-and-after photos on her Instagram page. Along with the seven million shares the post generated across Facebook, Twitter, and YouTube came a barrage of hateful remarks. "Be fuming if I woke up next to her the next morning," wrote one person. "Why I trust no bitch wearing makeup. They're covering up something," said another. "I'll spend some time with her, so long as I never have to see her before 10 am," wrote one more. Ashley's story gained more prominence because she bravely chose to make a YouTube video discussing the phenomenon. But this kind of mob mania happens daily, creating anguish for those against whom it is directed.

Mental Torture

The abuse of people's rights online is now manifesting itself as a mental health crisis among young people. Data from the U.S. Centers for Disease Control and Prevention (CDC) shows that between 2007 and 2021, the period in which social media and smartphones became ubiquitous in our lives, suicides rose 62 percent among those ages ten to twenty-four, whereas the rate had stayed steady between 2001 and 2007. A different CDC survey of high school students revealed that 22 percent had "seriously considered suicide" in 2021, up from 16 percent in 2011, with 18 percent saying they'd made a suicide plan and 10 percent saying they'd attempted suicide at least once, compared with 13 percent and 8 percent, respectively, ten years earlier. The numbers are even more alarming for high school girls, of whom 30 percent said they'd seriously considered taking their lives. More than half of the girls surveyed (57 percent) said they experience persistent sadness or hopelessness, up from 36 percent in 2011.

A statistical shift of this magnitude cannot be an anomaly. So, what is the common factor? Well, we think the U.S. surgeon general, Dr. Vivek Murthy, *almost* got it when his office noted in a 2023 advisory that a "growing body of research" showed that social media, used by 95 percent of all adolescents on a daily basis, has "potential harms." The surgeon general urged that we "increase our collective understanding of the risks associated with social media use, and urgently take action to create safe and healthy digital environments that minimize harm and safeguard children's and adolescents' mental health and well-being during critical stages of development."

Jonathan Haidt, a social psychologist whose books have done much to crystallize understanding of how social media has shaped our thoughts and actions, was disappointed with the surgeon general's statement. The time has passed, he argued, for equivocating on "potential harms." Haidt offered one explanation for why, despite the very clear correlations in aggregate mental health data, some experts suggest the jury is still out on all this: It's that empirical findings tend to focus on the direct, causal "dose-response effect" on an individual person from exposure to social media, whereas the phenomenon should be understood in terms of the wider, societal "network effect." Rather than studying social media as if it were, say, sugar or cocaine, examining in isolation the impact on a single brain, Haidt urged, we should think about how this plays out for teenagers whose entire social experience is largely lived online. In a February 22, 2023 blog post reflecting on the CDC's 2011–2021 study, Haidt offered the hypothetical examples of a twelve-year-old girl in 2011 whose parents give her an iPhone 4 when most of her friends aren't yet online and a twelve-year-old girl in 2015, by which time smartphone and social media use have become widespread.

Imagine, Haidt wrote, that in 2011, "the girl spending 5 hours a day on Instagram finds her mental health declining, but her friends' mental health is unchanged. We find a clear dose-response effect. If she were to quit Instagram, would her mental health improve? Yes." But fast-forward to 2015, "when most girls are on Instagram and all teens are spending far less time with their friends in person . . . [and then, at that time] a 12-year-old girl decided to quit all social media platforms. Would her mental health improve? Not necessarily." By then, if all her friends continue to spend five hours a day on the various platforms, "she'd find it difficult to stay in touch with them," Haidt wrote. "She'd be out of the loop and socially isolated. If the isolation effect is larger than the dose-response effect, then her mental health might even get worse. When we look across thousands of girls, we might find no strong or clear correlation between time on social media and level of mental disorder. We might even find that the nonusers are more depressed and anxious than the moderate users."

In a powerful statement of why we all must act together to fix the mental health crisis caused by our dependence on these data-controlling social media platforms, Haidt then summarized the problem: "What we see in this second case is that social media creates a *cohort effect:* something that happened to a whole cohort of young people, including those who don't use social media. It also creates a trap—a collective action problem—for girls and for parents. Each girl might be worse off quitting Instagram even though all girls would be better off if everyone quit."

Poor analysis of social media's harms plays into the platforms' legal defense. "We can't be policed for what bad people say on our platforms," they say, often citing the controversial Section 230 of Title 47 of the United States Code, enacted as part of the Communications Decency Act of 1996, which with the intent of enabling free speech exempted internet platforms from the kind of liability that, say, news publishers face for defamation and other legal challenges. "We didn't make Walker Farriel Montgomery kill himself. A scammer did," they would say. They'll also contend that they pay vast sums of money to monitor, moderate, and remove content whenever it's found to breach their internal policies prohibiting hate speech and socially harmful disinformation. (What they avoid saying is that they outsource this moderating task to sweatshop-like places in Africa and Asia, where low-paid staff must pore through the worst of what humanity has to offer and, as a result, often suffer from mental trauma themselves.)

But, if we extrapolate from Haidt's useful framing, social media companies' responsibility for the mental health crisis lies not solely in specific content that induces people to harm themselves or others but also in how their algorithms create an addictive dependence on a toxic social media environment. It's impossible to separate Walker's and other victims' drastic decisions to end their lives from the wider social media context, with all the cyberbullying and other forms of peer pressure that it encourages. Recognizing it from a holistic perspective—that the pain we're all witnessing is part of an ugly overarching system—is the starting point for understanding how we've been stripped of our rights and how to restore them.

The pernicious way in which online platforms allow their curation algorithms to encourage antisocial behavior that erodes people's self-esteem is, we believe, a form of mental torture. And, as with various forms of physical torture banned by the Geneva convention, it amounts to a breach of human rights. These systems are violating the sanctity of the person.

Asserting this requires a demonstration that some event in the digital realm directly led to a flesh-and-blood person's suffering in the offline world. If you accept our and Haidt's analysis in the preceding paragraphs, you'll recognize that evidence of a causal link is often incontrovertible. But if you still harbor any doubt that our real-world existence can be harmed by our online experiences, we now ask you to go one step further, to recognize, as we mentioned previously, that software companies are quite literally digitizing our flesh-and-blood identity. Your biological DNA (in the form of codified genetic information) and your social DNA (as represented by records of your online behavior) are both captured in digital form and stored in centralized data servers. This data is obtained by companies such as 23andMe and Ancestry, which exploit the connections that are easily made between these two versions of you. They find all sorts of ways to profit from them that you're not privy to. Your data is also exposed to the attendant risks that come with putting sensitive information online: In October 2023, 23andMe informed customers that hackers had breached its "DNA Relatives" feature, which allows participants to compare their information with that of other participants worldwide. Who knows who now has access to those customers' vital, highly personal information?

To be sure, the digitization of human biological information brings powerful potential benefits to society. If we start with an idea with which I'm pretty certain we all agree—that we own our biological DNA—and add to that our contention that we should own the digital version too, we can foresee a world where it can be put to positive use. Imagine the important medical breakthroughs that will be possible if we can willingly share such information in a manner that we control. Sadly, the biological data grab by Silicon Valley titans is only secondarily motivated by positive outcomes for society. It is primarily a move to shore up a dominant position in their capacity to manipulate how we communicate with one another.

One manifestation of the drive to shape our social behavior is found in what we might call the Instagram aesthetic. Scroll through TikTok or randomly explore a selection of Instagram "influencers"—a role that 57 percent of Americans ages thirteen to twenty-six said they aspire to, according to a 2023 Morning Consult poll—and you'll notice how the images tend toward conformity, even if each person posting is supposedly seeking to stand out from the crowd. It's the same bikini-clad pose on vacation, the same Kardashian-inspired pout, the same buff body in the infinity pool framed by the sunset, the same dude who's "crushing it" with his chin-ups. And now, with Magic Editor on Google's Android phones, everyone can modify their smiles and backgrounds to come up with a similarly idealized look. The internet idea many of us bought into during the early 1990s was that of a network that gave every person—each with their unique voice, appearance, and expression—an input into the great pool of ideas, and that out of that rich soup of ingenuity would spring the most amazing inventions and creations. But instead we are systematically stripping people of their individuality, of their capacity for self-actualization, of their authenticity. This doesn't feel very aligned with the American Dream to us.

By the way, if you're over fifty-five, we seriously hope the preceding paragraphs didn't elicit a smugness about the younger generation's image obsessions. Maybe you've never posed for an Instagram selfie on a Greek island or posted some shallow aphorism about your personal goals, but there's an almost 100 percent chance, regardless, that the algorithms have got you. We say this as two people over fifty-five speaking from experience. Look, between us, Michael and I have six Millennial and Gen-Z children. I also have three little ones from the latest cohort, which has been called Generation Alpha. None of our children have known anything other than life with the internet, at least from their teenage or earlier years onward. And when we talk to them, there's a savvy there, a wisdom—sometimes even a cynicism—about what can and cannot be trusted online. The reality, also, is that those who've fallen for the most extreme conspiracy theories online and converted those beliefs into political action, those who've embraced the most hard-line positions that internet pundits on either side of politics keep throwing at us—those people, predominantly, are older. If you ever feel like some other force is controlling your life, that you've lost control over part or all of it, it's because you have. We are not pointing fingers at any particular cohort. Our point here is that this is not a generational thing. The mission to reclaim our humanity sits with all of us, regardless of age, creed, or race.

The Digital Hypodermic Needle

How did we just let all this happen? One problem is that we've framed our rights in the internet age more as a privacy issue than as a matter of our personhood. People often say, "Sure, I don't like that these platforms are snooping on me. But I really have nothing to hide, so having them know stuff about me is a small price to pay for the convenience of what they provide." To us, that seems like a cop-out.

Here's a thought experiment I recently ran with a friend who gave me that "Privacy? Who cares?" line. It goes like this: What if the U.S. Postal Service came to you and asked, "Hey, would you like us to give you free stamps for the rest of your life?"

Perhaps you would reply, "Sounds interesting. Tell me more."

The post office would then lay out some conditions that come with the deal: "We will open all your mail. We will keep records of every person you ever communicated with, every business you ever transacted with, every magazine or book you ever purchased. We will install cameras and other monitoring systems in every room of your house, including your bedroom, as well as in your car, and we will use all the information we gather from them to decide what among our product offerings, and those of our clients, to bombard you with pitches for. We will flood your letterbox with junk mail to the point that it is constantly overflowing. Oh, and after secretly reading your thirteen-year-old daughter's diary, we will send her weight-loss brochures that actively use body-shaming methods to drive her to anorexia."

Would you take a deal of this kind? Free stamps in return for around-the-clock surveillance, unsolicited interruptions of your life, outright exploitation of all your information, and driving your children to self-harm? Of course, you wouldn't. So why do you accept this kind of predatory behavior and an even greater level of interference and surveillance in your online life today?

We've built a system that's dehumanizing individuals. Proprietary algorithms, trained by the logic of their corporate owners' profit motives, corral us into audience groups to which their clients can more readily sell goods, services, and ideas. They succeed at this—or at least succeed frequently enough to justify the fees charged to advertising clients, data brokers, and other players that monetize and exploit this information—by delivering stimuli that will routinely trigger chemical reactions in our brains.

An important aside: It's best to think of these algorithms as hybrid constructs. One part is pure math—the software code that tells an internet application to ingest the digital signals it receives from our online clicks and then determine which content to deliver to us. The other part is pure biology—the internal chemistry that drives our personal response to the stimuli on our screens, leading to the actions that generate the clicks. Finally, as per Haidt's analysis, all of that is wrapped up in the fact that we are social beings: We act in relation to others, and we have a chemically wired desire to belong and to gain recognition.

In the past decade, scientists have discovered a dedicated pleasure-pain circuit in our brain they call the reward pathway. Dopamine, produced by the adrenal gland, is the granddaddy of the neurotransmitters involved. In rewarding you with an innate sense of pleasure, while also administering feelings of pain, dopamine is charged with bringing the brain into balance—the biological state of equilibrium known as homeostasis. In regulating how we respond to things that make us happy and those that make us sad, fearful, or angry, it is supposed to keep us in check, managing activities like sleep, attention, learning, behavior, cognition, and even, for new mothers, lactation. But it's in the regulation of our so-called fight-or-flight response, a primal instinct that made sure our Stone Age ancestors instantly ran away at the sight of an oncoming lion or sought to protect a vulnerable newborn, that it creates complications in aspects of modern life. This chemical mechanism wasn't designed for the abundance of stimuli to which our brains are now subjected, which is why it keeps driving people to do harmful things to themselves. In the face of all this, our government institutions seem incapable of addressing it. The famed sociobiologist E. O. Wilson said it well in 2009: "The real problem of humanity is the following: We have Paleolithic emotions, medieval institutions, and godlike technology."

Dopamine releases and their effect on behavior are a numbers game. The more reinforcing a certain stimulus response is to the reward pathway, the more dopamine is released. The Stanford psychiatry professor Anna Lembke uses the colorful analogy of a seesaw in a children's playground to explain how, when we experience something pleasurable, the chemicals associated with that pleasure drive one side of the seesaw down, which then prompts a bunch of "pain gremlins" to jump on the other side to bring things back into balance. "That's that moment of wanting a second piece of chocolate or wanting to watch one more TikTok video," Lembke said during one of her entertaining presentations. "Now if we wait long enough, those gremlins hop off and homeostasis is restored. But what if we don't wait? What if we continue to ingest our drug of choice? Over days to weeks to months to years, those gremlins start to multiply and pretty soon we have enough gremlins on the pain side of the balance to fill this whole room. . . . This is what the addicted brain looks like. Those gremlins are now camped out on the pain side of our balance. . . . Now we need our drug of choice not to feel pleasure, but just to level the balance and stop feeling pain."

This should be familiar to anyone with any experience, either directly or through a loved one, of narcotics or food addictions. Plenty of research demonstrates that large amounts of dopamine are released with the intake of cocaine or amphetamines. What's now clear is that the same stimulus-dopamine-pleasure-pain feedback loop occurs with visual, audio, and other non-ingested stimuli. It's why Lembke, who has spent most of her career studying drug addiction, has now turned her attention to what she sees as an even more powerful addictive force than hard drugs: social media, the ultimate driver of abundant, out-of-control stimuli. She calls the smartphone the "modern-day hypodermic needle," the delivery mechanism to which we now turn for whatever it is that we crave, whether it's validation, distraction from boredom, or a trigger for our ever-present propensity for anger.

Through their mathematical, probabilistic logic, social media algorithms have long tapped into this chemical mechanism so as to keep us coming back to that computerized hypodermic needle at our fingertips. And, according to Facebook's first president, Sean Parker, this was all by design. Speaking at an Axios event in 2017, Parker said he and the other early builders of the platform knew that if they gave "you a little dopamine hit every once in a while because someone liked or commented on a photo or a post" that would "get you to contribute more content and that [was] going to get you, you know, more likes and comments." This "social validation feedback loop," he said, is "exactly the kind of thing that a hacker like myself would come up with because you're exploiting a vulnerability in human psychology. I think that the inventors, creators—it's me, it's Mark [Zuckerberg], it's Kevin Systrom at Instagram, it's all these people—understood this consciously and we did it anyway." That same year, Facebook's former head of user growth, Chamath Palihapitiya, expressed "tremendous guilt" over what he and others had built. He told a Stanford Business School conference that they "kind of knew something bad could happen" and that they "created tools that are ripping apart the social fabric of how society functions."

Beyond these admissions, there has been precious little accountability from the platforms for the harms they have done. This is despite a modest but growing legal effort against them. In 2022, a UK coroner delivered an important finding on the suicide of Molly Russell, stating that the fourteen-year-old London girl "died from an act of self-harm while suffering from depression and the negative effects of online content" after viewing a large amount of content on Instagram and Pinterest related to suicide, depression, self-harm, and anxiety. Her grieving father put it perfectly: Meta and other social media companies, he said, were "monetizing misery."

Elsewhere, two bombshell developments just before this book went to print revealed just how little has changed. Misaligned financial incentives continue to encourage a handful of corporate elites to exercise their excessive power over our data to put human beings at risk. In one of these cases, a lawsuit filed by thirty-three state attorneys general provided evidence that Meta suppressed proof of the mental harm Facebook and Instagram had done to youth. The AGs said the company ignored staff concerns about a Big Tobacco–like denial of that harm because it was unwilling to give up young users' high "lifetime value" (a measure of how much revenue a customer is expected to generate over their life span). Relinquishing that opportunity would have resulted in a 1 percent hit to revenue—tiny in the grand scheme of things but too big, apparently, for Meta's top brass to relinquish in favor of doing the right thing for children.

In the second news event, the dramatic ouster of OpenAI's CEO, Sam Altman, and his swift reinstatement to power in a matter of days in November 2023 exposed how board members' concerns over the "safe" development of the company's generative pre-trained transformer (GPT) technology ultimately lost out to the higher financial returns that Altman's leadership promised. We'll explore this case more deeply in chapter 6; for now, it's sufficient to point out that it is part of a pattern of behavior we've noted in Silicon Valley. There, many more prominent voices will often describe the dangers of creating artificial general intelligence (AGI)—software that can think like a human, with the ability to self-teach—but, then, in the next breath, will say, "Trust us. We'll fix this for you." These are the same people in the Bay Area who, reports say, won't let their own kids use social media.

There's a clear parallel here with the opioid epidemic, which for a long time wasn't recognized for the crisis that it was or addressed with the urgency it required. The discussion was about people having the right to treat their pain with new drugs, such as Purdue Pharma's OxyContin, prescribed by doctors, not about the vicious dopamine cycle of stimulus-response that users were grappling with. The narrative made it hard to notice how entire communities were being ripped apart by overdose deaths, crime, and family breakdowns.

I have some experience with this. When I was president of the South Boston Neighborhood House in the 1990s, we were confronted with an inexplicable and unprecedented suicide cluster in the community. Young people, in large numbers, were taking their own lives. This prompted us to open our doors around the clock to address the immediate needs of the community. Working together with the other nongovernmental organizations in this tight-knit South Boston community, we also went to work to determine the hidden cause of this tragedy. With the help of experts, we learned over time that large segments of the population had become addicted to OxyContin. Many of these individuals eventually gravitated to heroin in search of an even faster, more intense opioid release. This accelerating downward spiral, accompanied by a complete loss of hope, led many, sadly, to end their lives. Only when nearly four hundred young people who had lost family members and friends gathered in a church and decided to take matters into their own hands did the crisis come to an end.

Now, with our social media addiction, our "digital opioid" problem, we need to act much like these young people in South Boston. We must recognize and admit that we have a problem, own it as a community, and take action together. Why, after so much evidence that the companies have been aware of the harm done by their algorithms—including the explosive 2021 testimony from Facebook whistleblower and Project Liberty collaborator Frances Haugen—have actions not been taken against them such as those that shut down Purdue Pharma?

A significant blocker when it comes to consensus on how to resolve the harms caused by social media is the forum of discussion: Debates over these issues, as with arguments on most issues of importance today, typically occur within the toxic, mind-manipulating digital environment in question. It becomes quite circular: Our arguments about digital platforms occur on those same platforms, which are deliberately designed to get us to argue and avoid consensus. In their bid to keep us scrolling and engaging with their services for longer periods, social media feeds will favor posts that evoke strong emotional responses, of which anger is one of the most powerful. When combined with the audience-clustering patterns perfected by Facebook, the algorithms corral us into echo chambers as the dopamine-release function prompts people to "like" one another's contributions, giving validation to strong opinions. Thus, like a bunch of addicts encouraging one another to snort more and more lines of cocaine, the members of a politically like-minded group will reinforce one another's anger-stirring instincts. The result: a vicious cycle of outrage and polarization as increasingly triggered groups of people who share the same opinions angrily face off with opponents on the other side of some divisive issue. Nuance and detail are cast aside, as is the prospect of thoughtful compromise. It's why we can't agree on how to fix social media—or on anything, for that matter.

When news emerges that bad actors, including state actors, have been using social media to sow discord and polarization in U.S. society, the blame often gets attributed solely to those actors. And, sure, in the case of Russian interference in the 2016 presidential election and in Meta's move in 2023 to shut down thousands of fake Chinese accounts that had been inserting divisive content into Facebook groups, we find evidence of these players' actions. But we often miss the bigger, more important point: that we, by submitting ourselves to this system of surreptitious manipulation, have ourselves unwittingly become the agents of our own social dysfunction.

Again, some of you may be thinking, "This isn't me. I'm in control." That may be the case, but, if so, research is showing you're an outlier. Our addiction to social media is making it very difficult for us to live without our smartphones. Eleven studies from experiments reported by *Science* magazine found that "participants typically did not enjoy spending 6 to 15 minutes in a room by themselves with nothing to do but think, that they enjoyed doing mundane external activities much more, and that many preferred to administer electric shocks to themselves instead of being left alone with their thoughts." You could also do your own experiment: Try taking your kid's smartphone away and see what happens.

Maybe we've always hated being alone. Maybe we've always struggled to avoid boredom. But we've never before had such a readily available instrument that can deliver mindlessly addictive drivel to address it. The void that exists when we are alone represents a rich opportunity for social media services. By filling it, they can grab something that every business wants to capture in the digital economy: some of your limited time and attention.

Who among us wants this? Who's proud of the fact that our average daily time spent on social media went from 147 minutes in 2021 to 151 minutes in 2022? If you're thinking 4 minutes isn't much, try multiplying that daily number by the five billion people who now use social media worldwide and then annualizing it. That's an extra 7.3 trillion minutes—the equivalent of 5.1 billion human days—that these companies sucked out of our lives in 2022, and that they converted into advertising dollars. We have a serious, global addiction problem, people. This is a public health threat that overshadows even the recent global pandemic. We must address it.

The New Human Right

There's a way out of this mess. We promise we'll get to it. But before we outline our proposal for a next generation of the internet that preserves people's rights as individuals, it's worth going one more layer down on what we mean by the word "rights." The

Bill of Rights and other texts it inspired, such as the UN's Universal Declaration of Human Rights, offer one way to look at the matter: a laundry list of distinct areas of human endeavor that individuals should be free to pursue without the interference of the state. Such documents play a vital, functional role in democracy by giving the weight of law to a set of explicit, clearly defined rights that cannot be violated. But they're less vocal on the bigger purpose to all this, and that's the notion captured in the Declaration of Independence's phrase "the pursuit of happiness." We must understand the "why" behind the laundry list. And in our reading, that "why" boils down to the idea that each of us deserves the opportunity to try to make something of the life we've been given and to do so on *our* terms. It speaks to the capacity to self-actualize, to develop an identity that is ours to express, not that of some group into which we've been coerced by the nation-state or by some other power, such as an internet algorithm, that's sapped us of our free will. It's about the right to be and develop your genuine *you*. That's the standard we need the new internet to uphold.

Two decades ago, during his naturalization ceremony in 2003, my coauthor, Michael, experienced an epiphany around this concept. It was a moment that taught him a valuable lesson on the meaning of the American Dream. Here we've included a brief passage from a longer essay he wrote about the experience:

> Sitting on folding chairs with hundreds of other new citizens inside a big tent installed in the parking lot of an Immigration and Naturalization Service facility in Garden City, Long Island, I was forced to grapple with my inner snob during the opening, seemingly cheesy moments of

the ceremony. The facility's general manager, a portly gentleman wearing a Stars and Stripes tie, introduced various members of his staff seated behind him, describing the selection of his coworkers as "dignitaries," each of whom stood up to receive a round of applause. They weren't exactly eminent persons, so the label seemed excessive. My journalism career had frequently put me in touch with political and business leaders at the top of the power pyramid, so this, in contrast, seemed to my self-absorbed brain to emphasize how distinctly lowbrow this ritual of becoming an American was shaping up to be.

But, then, after we'd recited the Oath of Allegiance, something magical happened. "And now," declared the general manager, "Joan from application processing will lead us in the national anthem." (In truth, I don't remember her name or title. I only recall that this woman occupied some kind of routine office role.) "Joan" took to the floor and, without backing music, started belting out, not the "Star-Spangled Banner" standard, but the gospel-inspired version made famous by Whitney Houston. As she took it to its rousing crescendo, the entire crowd before her got to their feet. Whooping and cheering while enthusiastically waving the mini Stars and Stripes flags they'd been handed, hundreds of new Americans raucously willed Joan to the song's heart-stopping finale.

It was an unbridled celebration of humanity, and it shamed my inner snob into silence. I now understood what this public servant had artfully done, gaudy tie and all. He'd given us a demonstration of what the American Dream represented: that in this country, everyone, whether

> or not they ever get to be called a "dignitary," is dignified. That dignity stems from them having as much right as anyone to pursue their dreams, whether it's to work for a government office or to sing on a stage with all their heart. I also learned something from those I was sitting amongst that day: that as Americans, we wholeheartedly cheer on one another's efforts to live out this idea, that in embracing it for others we affirm it as a possibility for ourselves. It was the perfect message with which to start an American life.

Europeans, Asians, Africans, and Latin Americans don't typically migrate to the United States with a desire to conform entirely to some universal standard of Americanness. They come here because they see it as the best place for people to try to be who they want to be, embracing both the culture of the place they left and that of the one to which they moved to create something new in their lives. So, with that idea of the American Project in mind—a system that celebrates people for being their authentic selves, not because they meet some externally ordained, inaccessible standard of beauty or perfection—we assert here that the new internet must protect and preserve a human right to fulfill a similar objective. It's not a right that guarantees success, promises a handout, or gives assurances that we won't fail or suffer. It's simply the right to give each of us a fair shot to attain our dreams and that some other external power can't constrain or shape that effort against our will.

We call this a right to self-actualization. And when it comes to the internet, this right cannot be attained without first securing control over our data. That's why the one big takeaway to retain from this book is that we, the human beings who use the internet, have a *right* to our online data and to not have it used against us by others. Within the power dynamic that shapes our lives in the internet age, where our digital and offline worlds have become one and the same, that data is a direct expression of you. Control over it defines whether you or someone else has control over your life. Right now, someone else is the gatekeeper for your data and you have no idea what they know and how they are using it. In effect, they control *you.* Does that seem fair? Don't you think you have a right to say how it is used? Shouldn't it be defined as *your* data, under *your* control? Shouldn't *you* own you?

The internet platforms are abusing our rights. They have trampled on both our right to agency and our right to privacy. They've denied us a property right over the data and content we generate, one that will become even more flagrantly abused as Big Tech starts converting its vast internet data stores into profit-making generative AI products.

Yet society has not sufficiently woken up to this. So, taking our cues from the Enlightenment philosophers cited in chapter 1, it's up to those of us who see the need for change to assert a claim to these new rights even before they are universally recognized by society or the law. We must forcefully make the case for this new mindset, because we fully expect that any demand that the platforms give us control over how our data is used will be met by an army of lawyers brandishing the terms and conditions contracts to which people routinely click "accept." Let's not get bogged down by legalese that gives the platforms carte blanche to do with us as they please. We can make an obvious and overwhelming *moral* case that those "terms and conditions" contracts are voided by a higher law respecting the sanctity of the human being, one that's enshrined in the U.S. Constitution and in the UN's Universal Declaration of Human Rights. There are plenty of examples of laws and contracts being nullified on these grounds. Consider how the American Project, over time, has rectified many of the injustices it had at first accommodated. All but the most extreme white supremacist or misogynist would now vehemently reject the idea that another human being should be treated as property or as unequal. Rights previously denied to African Americans and women are now enshrined in amendments to the Constitution. No piece of paper can override them.

Tristan Harris and Aza Raskin of the Center for Humane Technology have observed that any transformative new technology will give rise to a new class of rights. "We didn't need the right to be forgotten until computers could remember us forever," they note, "and we didn't need the right to privacy in our laws until cameras were mass-produced." That's precisely what has happened with the all-encompassing data economy that the current internet and its related technological inventions have created. We need our moral and legal concepts to catch up.

Decentralized to Centralized

To achieve meaningful change, it is not enough to just update the law. We must embed our rights directly into the technology, making them functionally executable. And for that we must add new protocols and systems on top of the half-century-old base-layer protocols and systems on which the internet is based. To comprehend why such a core change is needed, we must first examine how the internet was initially architected and how it evolved to where we are now.

In 1969, something called the Advanced Research Projects

Agency Network (ARPANET) was launched, marking what is commonly referred to as the beginning of the internet, its first generation. Initially connecting the computing departments of four California universities, ARPANET was funded by a military agency that today is known as the Defense Advanced Research Projects Agency (DARPA). There were various motivations for building this decentralized network, but a major one was that, over the preceding decade, U.S. military officials had become worried about a Soviet threat to the national telecommunications system, which was at that time managed by telephone switching centers. They worried the system's dependence on those centralized choke points would render it inoperable in the event of a Soviet nuclear attack. The idea behind ARPANET was to create an alternative to that network. If computers could communicate without routing information through centralized hubs, there would always be a way to deliver messages, no switching hubs required. Uppermost among the internet funding agency's concerns, then, were not the interests of the ordinary people who would later use it, but the government's Cold War fears.

Regardless of whether and how human beings factored into the internet founders' early designs, there was no getting around the fact that the system's foundational protocols—a group of common rule sets enforced by shared software to which independent computer users implicitly agree—had to address the identity of computers, not of people. This machine-centric design was integral to an all-important, intertwined pair of protocols, one called the transmission control protocol (TCP), the other the internet protocol (IP), together better known by their combined abbreviation, TCP/IP. The code for this pair was developed by Vint Cerf and Bob Kahn in the 1970s and formally launched in

1983. The adoption of this new combined protocol for efficiently sharing packets of data enabled a much wider network than the original ARPANET. Starting with the notion that authorized users would need unique identities, the system enabled internet service providers such as cable companies to assign each participating computer with something called an IP address. To this day, that term refers to the special number your device uses when it logs on to the internet to communicate with outside servers and computers. (Note: It logs on to the internet; you don't.) If you are ever having connectivity issues, it might be because the network doesn't recognize your laptop or smartphone's IP address. It's about the ID of the device, not the ID of the human user.

For our purposes, it's useful to think of a core protocol like TCP/IP as something akin to the U.S. Constitution: It represents an especially thin piece of code, a set of basic, inviolable principles to which the pieces of code (or laws) built "on top" of it are bound. So, just as the Constitution, ratified in 1789, enabled the vast array of U.S. laws enacted at federal, state, and municipal levels in the centuries that followed, so too did TCP/IP, put into place in 1983, unleash a host of new computing arrangements to enable different types of network activity.

Through the 1980s, these "up the stack" enhancements were mostly concerned with different ways to connect computers with each other in direct, *one-to-one* exchanges of information. New higher-level protocols were developed to enable email, for example, and file transfers. Then, in 1989, on the two-hundredth anniversary of the U.S. Constitution, an Englishman came up with a new protocol that directly gave people access to all manner of information broadcast over the internet. That model birthed a new era in which the internet would generate massive amounts of human data—and we mean massive. This was Tim Berners-Lee's hypertext transfer protocol (HTTP), which along with a new coding language and something he dubbed the uniform resource locator (URL), launched the World Wide Web. This marked the beginning of the internet's second generation, the one most of us think of when we refer to the internet: the web. (Dave Clark, an MIT professor and early internet inventor who's now an adviser to Project Liberty, argues that we've since left the web era and are now in the "app age.")

Soon the internet would bear little resemblance to the DARPA network of the late 1960s. Spurred on by increasingly easy-to-use browsers, from Mosaic and Netscape to Microsoft's Internet Explorer to Google's Chrome, the web's endless pool of information would draw in millions and, eventually, billions of people. As the people came to it, they brought with them their combined attention, delivering it en masse to single-location entities known as websites. They also brought their data with them. It was as if each website was an event site in which the owners threw a party we could all attend. The problem is that we had no idea we were leaving behind the keys to our houses when we walked in the door and joined the party.

In the 1990s, once businesses realized the web's potential for e-commerce, a couple of thorny issues arose: There was no system for authorizing this swarm of humans and businesses that now wanted to transact with one another. It wasn't something that the base-layer IP address system for computers could deal with. The solution had to come at a higher layer in the internet's software and protocol stack. And, as was long understood by Dave Clark and other internet founders—who resisted government security agencies' efforts to include human information into the data shared over TCP/IP—creating a common database of human IDs would have been a recipe for domination and abuse by whoever controlled that information. Yet without a universal identification system, how could people and businesses be trusted to make payments or to fulfill promises to deliver goods and services? In time, jury-rigged solutions emerged. A few companies attained vast certification powers by default, simply on account of the large database of users they controlled. These entities amassed outsize, government-like power. They established themselves as the internet's gatekeepers, sowing the seeds of an even bigger problem in the new millennium. This imperfect solution forged the autocratic, centralized surveillance system within which we now toil.

One way to think about this era is that it was one of centralization. The internet was initially intended as a decentralized network, with no hub in charge—and the base-layer protocols have stayed that way. This model was in harmony with the decentralizing spirit of American democracy, which was designed to prevent the accumulation of power by one person, body, or branch of government. Post-1989, however, when the World Wide Web's runaway success drew humankind (and its data) onto the internet, huge gatekeeping powers started accruing to companies that accumulated that bounty. Much like the very model that America's decentralized democracy replaced, in which a king could dictate what happened to his subjects, these new centralized internet entities became outrageously powerful.

By the early 2000s, it became apparent to Silicon Valley's savviest investors that among the companies best placed to become a king of the internet was one with the quirky name of Google. The company's powerful algorithm (a form of AI), which perpetually ingested and analyzed search information supplied by users, had successfully resolved the chaos and disorder of the internet. Where companies such as Yahoo!, Alta Vista, and Ask Jeeves had tried less successfully to solve search through a cataloging model, Google took a different approach: It learned from us, establishing a ranking system that created a prioritization model based on what people were looking for. It was a machine that would become ever more efficient and precise and so deliver an ever-more-satisfying search experience. Google quickly came to dominate the search business.

Just as important to Google's investors, the company had figured out how to turn that search dominance into a money machine for online advertising. Inventions in the 1990s such as Netscape's tracking cookies and DoubleClick's click-through banner ads had promised a gold mine for companies seeking to convince people to buy things at websites. But it wasn't until a Google team discovered how to monetize the company's ever-growing pool of people's search data that the online ad business really took off. In the process, they launched the business model that Shoshana Zuboff later dubbed "surveillance capitalism." The Google team figured that the combination of search terms, cookies, and IP addresses that were feeding into its search algorithm would help advertisers more directly target customers. The data-driven approach was later copied by virtually all other internet platforms. Many a Silicon Valley pitch deck essentially touted the same model: A company would build a massive audience by delivering "free" services over a platform whose network-effect appeal would make it indispensable, extract data from all those unsuspecting users, and then run ads tailored to them.

The company that most aggressively ran with this model was

Facebook, which recruited members from the Google team in 2008. Its "breakthrough" was the formation of "like audiences." These groupings of similarly minded people, so named for the commonalities in the posts their members rewarded with Facebook's thumbs-up "like" button, became ideal targets for advertisers. Their tastes and inclinations could be gleaned from how they interacted. It became clear that the best way to reinforce the stickiness and marketability of these groups was to feed them content with which their members would be inclined to engage. Others followed suit: YouTube's recommendation algorithm was one manifestation; Twitter, later rebranded as X, eventually adopted something similar with its "For You" selections. And so the far-reaching implications of Berners-Lee's invention were realized, but probably not in a way he had imagined. The giant intermediating, for-profit platforms had co-opted the internet. It's how they came to own the world's data. It's how they came to own us.

A Proposal: DSNP

Let's recap: The first generation of the internet, enabled by the protocol pair TCP/IP, was about connecting machines to the network. The second, referred to as the World Wide Web and enabled by another protocol, HTTP, gave rise to an era of mass data aggregation. We must now begin a third generation, an internet that safely connects people. In effect, it's about *re*-decentralizing. To achieve that, we propose doing something akin to what Americans have done at various times in the nation's history: amending the Constitution in order to enshrine newly recognized rights or obligations that had previously been neglected. Much like its constitutional equivalent, we need an amendment to the internet's protocols.

The solution we have in mind is yet another protocol, something we call the Decentralized Social Networking Protocol (DSNP). The brainchild of Braxton Woodham and Harry Evans, and backed by Project Liberty, DSNP puts you in charge of you by returning control of your data to its rightful place. By adopting this protocol, individuals would regain authority to decide who gets to see different types of information about them and their social connections. DSNP would be implemented on top of TCP/IP, adding a new credentials-proving layer that allows you to create and prove control over a self-identifier that functions as a kind of universal login. TCP/IP's transmission functions and HTTP's data linkage functions would continue. But with DSNP now added to the mix, you would engage with those protocols with exclusive authority over your identity and your data. Rather than surrendering that authority to whatever private company's platform you log on to, you would retain control.

Here's one vital, and very attractive, aspect of this new model: You won't be logging on to separate applications with multiple different usernames and passwords; you'll have one DSNP login. Of course, the connection to the internet itself will be physically established by your device, with its IP address. However, in this new age of the internet, your user experience will feel like *you* are the one who's logging on. The same will go for companies and businesses that wish to transact with you, with both parties choosing whether to grant access to information. The key point is that you will have much more control over vital data about you, including the network of connections known as your social graph and the activity you generate through those connections. Remember how we said big platforms became the gatekeepers of others' identity in the second, data-centric phase of the internet? Well, now you become the gatekeeper of your information, of who gets access to it and on what terms.

DSNP integrates a so-called decentralized identifier (DID), a software standard recently approved by the World Wide Web Consortium that can be viewed as a universal human passport for the internet. It proves that you, and only you, are who you say you are and have access to information about you. DSNP uses a digital wallet as a tool to grant permission for sharing this information with others, be it to provide unique personal data such as a medical record or driver's license, to post something on a social media feed, to follow or like someone or something, or to make a payment. Whereas people tend to think of digital wallets as smartphone-based apps for managing money-like items such as digital currencies, credit cards, online banking, and so forth, we think of digital wallets as similar to the leather wallets we carry around in the analog world. Those physical wallets aren't only used to store banknotes, credit cards, and debit cards; they're also the container from which you extract your driver's license, your health insurance card, your employee ID card, and other credential-certifiers. Digital wallets will have a similarly wide set of use cases and the additional capability to store all kinds of other credentials and assets that will in the future be digitized.

When you install a single DSNP-enabled wallet app on your phone, laptop, or other device, it will function as both your store of personal information and your mechanism for controlling others' access to it. It will allow you to grant selective permission to third-party providers that require specific information about you. It puts you—not Google, not your bank, not your doctor, not the institution where you were educated, not all the other businesses that currently store information about you—in the driver's seat. Where those institutions may continue to attest to your credentials or other records, you'll use your wallet to authorize them to share such attestations with third-party operators in need of them.

Your DSNP-enabled wallet gives you the power to provide whatever range of information accessibility, from minimal to maximal, you feel is needed for different contexts. Imagine convincing a bartender they can lawfully serve you a drink without showing them your driver's license, with all its ancillary private information about your birthdate, address, gender, and so forth. In this new model, you parcel out the bare minimum of information needed: incontrovertible, certified proof that you can answer yes to the binary question of whether you are over the legal drinking age. Or imagine a social media community of close friends in which you share photos and tidbits of information about your family with others who do the same and can know without a doubt that neither the platform on which the group lives or its advertisers have any access to that information without your permission. Such is the restoration of control afforded to individuals and to the groups they willingly form when the right to grant access to personal data resides with the individual to whom it belongs.

With the help of advances in cryptography and record-keeping, this model will allow a human being to prove they are a unique person (i.e., not a bot or an account disguised as someone else) while still withholding personal identifying information such as their name, birthdate, biometric data, or address. We've discussed the Four Rs. Well, this is the Three Ps: *personhood, proofs*, and *privacy*. Nonhuman entities, whether they are bots or legitimate businesses, will also be able to control DSNP wallets, but only real flesh-and-blood people can associate uniquely human attributes and credentials with their identities. The ramifications for addressing the problems of anonymity and bot-based misinformation are vast. Imagine an internet where you must be a real person if you are going to represent yourself as one, where you can only be one person, and where machines can't masquerade as people. Compare this evolved version of the internet to the one we're stuck with now.

These concepts are often lumped under the topic of self-sovereign identity, which sounds like a colorful way to describe the individual *rights* aspect of what we're trying to achieve. Self-sovereignty suggests you are in control of your data and of how you self-identify, that *you,* not some external authority such as a government or a company, are "sovereign" over who you are. But the MIT professor Alex "Sandy" Pentland, a Project Liberty fellow, warns against overusing the term "self-sovereign" because, in his mind, it wrongly implies that individuals and their identities persist independently of the society in which they live. "There is no 'you' without the various communities to which you belong," Pentland says. "Your identity is defined in relation to other people."

~

Whether or not Pentland's campaign against the term "self-sovereign" succeeds, his point about the interdependence of individuals and the society to which they belong offers a good transition to the second of our Four Rs: *responsibilities.* The idea is that the individual rights we receive as citizens necessarily come with obligations to the wider society in which we live. We need a similar concept for the next generation of the internet. So, how do we construct a prosocial internet in which we are incentivized to develop useful connections and engage in healthy, collaborative interactions in fulfillment of our responsibilities to one another? The answer lies in a dramatically different model for an important concept known as the social graph, which we briefly referred to earlier. The next generation of the internet will come with a decentralized universal social graph. It will no longer be treated as a proprietary resource for centralized platforms to use in ways that have brought out our worst antisocial instincts.

That's the topic of chapter 3.

CHAPTER THREE

Responsibilities: The Social Contract in the Internet Age

Ever consider the wonder that is the youth soccer league network of the United States? Throughout the country, every weekend, girls and boys ages five through eighteen play soccer on recreational or travel teams. It's all volunteer-led, from the local and county-level organizations that keep track of scores, game schedules, and team standings in the various divisions, to the moms and dads who coach the teams, to organizers who collect dues and make sure the teams have jerseys, equipment, and well-tended fields, to the rotating "snack duty" assigned to a different family each week. Other than perhaps that the games are typically played on fields owned by the town or public school, there is no government involvement in the process. It all just kind of happens organically.

Volunteering is one way a functioning society converts mores, norms, and cultural affinities into positively reinforcing, prosocial behavior. But there are many other forms this takes, from simple things like waiting your turn at the DMV or giving up a seat to an elderly person on the bus, to participating in blood drives or medical trials, to donating to charitable organizations or public broadcasting services. Different cultures manifest prosocial practices differently. In Japan, visitors marvel at how lost-and-found boxes on subway platforms are left unlocked and unattended, with everything from smartphones and cash-laden wallets inside. In places with strong familial culture such as Argentina, people go out of their way to shower strangers' children with love and affection, and no one thinks for an instant that it's odd.

In democracies, these behaviors capture the society-focused *responsibilities* side of the citizenship bargain, the flip side of the individual-focused *rights* piece. The degree to which these behaviors are practiced is a measure of the strength and sustainability of a civil society. Together, they are the glue that holds communities together; they foster the bonds of trust with which we collectively protect "the commons," the set of cultural and natural resources that no one owns.

The commons concept dates back to the shared land and water resources such as pastures and streams that "commoners" in England were allowed to use—to graze cattle, for example, or to fish. More recently, its meaning has been extended to cover ideas and creative works in the public domain, cultural creations over which there is no copyright or private claim of intellectual property. The word "commons" can also refer to the idea of a town square, either a physical place or a virtual one, where people are free to openly express and debate ideas without interference from private or state power. In all these manifestations, it takes mutual respect—expressed in voluntary, prosocial behavior—for the commons to thrive. Each act of kindness or expression of empathy, courtesy, or appreciation, big or small, subtly reminds us that we share in the human experience and that, with the help of others, we can all better confront the challenges of life. By building similar instincts into our actions, those reminders create a positive feedback loop where kindness, empathy, and respect breed further kindness, empathy, and respect. By extension, society itself becomes stronger.

Those who fancy themselves rugged individualists might find these observations hokey. But the idea of people bonding in communities is deeply ingrained in the American psyche. It's why nineteenth-century stories of fearless men and women establishing new lives on the frontier live alongside contemporary tales of rural towns whose resilience is burnished by a communal spirit in which everyone looks out for one another.

Here's the thing, though: Social mores must be nurtured, cultivated, and protected. History is full of accounts of societies losing the bonds that once held them together. Extreme examples include Sarajevo and Rwanda, where outbreaks of ethnic violence in the 1990s turned once-friendly neighbors and classmates into violent adversaries. But the erosion of communal bonds can also happen so gradually as to be nearly imperceptible. Even as empathy-enforcing aspects of a culture will persist in some corners, other cultural practices might be eating away at civil society. In Tokyo, where the lost-and-found boxes are still going strong on subways, thousands of young girls are steered into a form of prostitution as Japan's entertainment industry normalizes the idea of adolescents as sex objects. In Argentina, where my coauthor, Michael, ran a news bureau for Dow Jones in the 2000s, a sociable atmosphere prevails—a friendly waiter once scooped up Michael's three-year-old daughter and bustled her off for a tour of the kitchen—yet Michael also found the country's highways to be lawless death traps where people drive with reckless abandon. And in the same U.S. communities as the soccer leagues? Well, just keep reading this book for all the evidence you need that people are losing their respect for others.

Exploring the breakdown of civil society can help us understand the conditions needed to protect it. It's vital that everyone who participates in a society can confidently act on the understanding that others are also going to do the right thing. To get it working correctly, you need all Four Rs—*rights, responsibilities, rewards,* and *rules*—working together. It's part of the bargain we agree to in an open democracy. It's what Rousseau, as we noted in chapter 1, called the "social contract."

The case of Argentina in particular shows how important trust is for building a civil society. The story of that resource-rich South American country, with its diverse, well-educated population, is a sad one of ongoing economic failure. After repeated political and monetary meltdowns throughout the twentieth century, it fell from its perch as the world's 7th richest country in 1903 to only the 140th richest in 2003. Painfully aware of that decline, locals will admit that it was accompanied by a loss of respect for public spaces and the common good. In one memory that became seared in his mind, Michael, out walking one sunny day in Palermo, a well-to-do neighborhood in Buenos Aires, observed an elderly man standing high up on a balcony of a luxury apartment building finishing his coffee and then throwing the empty polystyrene cup to the street below. Why? Surely he had a trash can in his apartment. More than mere neglect for the commons, the man seemed to be expressing outright contempt for it.

The image of the cup-tossing man points to a contradiction, for Argentines hold passionate views on how to improve society and are fanatically proud of their country, especially of its World Cup champion soccer team. Michael developed deep friendships there, with people who were warmer and more hospitable than those he'd known anywhere else. The country proved a fabulous place to raise young children. But he came to glean that, at the wider national level, Argentina's social contract—that unwritten agreement by which democratic societies define citizens' relationships with their government and with one another—had long ago evaporated. No one trusted the government to act as a steward of the common good, so no one felt any obligation to contribute to it themselves. Tax avoidance was rife, and wealth was stored in U.S.-dollar-denominated assets, not in assets denominated in the derided local currency, the peso. As such, almost like clockwork, the economy would succumb once every decade to twin fiscal and inflationary crises. And the country's politics? They're completely performative: Populists from the left rail against the excesses and corruption of populists on the right, who do the exact same back at their critics.

Argentina's presidential election of 2023 is a case in point. It was won by Javier Milei—an ultraconservative former rock musician turned libertarian economist who promised to end the central bank and exhibited authoritarian instincts—mostly because Argentines were sick of the inflation- and debt-riddled mess in which the previous government had left the economy. History suggests that Milei's policies, which include returning the monetary policy to the same rigid dollar-backed system that got Argentina into trouble in the late 1990s, will also fail and that the Peronists will use their vast, corrupt political machine to claw their way back into power. The Argentine seat of government has always bounced back and forth this way, one corrupt, feckless administration after another. Argentines are wonderful, but their society is broken.

Sounds eerily familiar, right? It's as if the breakdown in Argentina's social contract is a harbinger of what successful societies like the United States face in the social media age. The corrupt, exploitative systems that govern people's lives have deliberately stoked performative tribalism, resolving nothing. The result is widespread mutual mistrust and a breakdown in the social contract. The experience of Argentina, which in 2023 suffered yet another currency devaluation amid triple-digit inflation, should be a warning to us all.

Yet the main positive aspect of Argentina, the familial warmth that people show to one another in the right settings, is also instructive. Michael and his wife, Alicia, would never have let that waiter carry their three-year-old off into the kitchen—no matter how much it made her squeal with delight—if they'd had reason to believe she would be neglected or, worse, abducted or abused. In the Argentine context, the same people who might drive like dangerous maniacs could be fully trusted to treat other people's children with kindness, respect, and care.

You can probably guess where this is going. Can we trust that the platforms that profit from social media chat rooms are doing so in ways that don't harm us? More specifically, to take the Argentine waiter analogy further, can we trust them not to harm our kids? No. Even if we're only now understanding the extent of the harm that has been done to our minds and those of our kids by participating in addictive, divisive online environments, we also know that the system is broken, that people, companies, and software platforms cannot be trusted. That breakdown in trust is a recipe for undermining the social contract, which, as the wider

Argentine societal experience shows, leads to a vicious feedback loop of bad behavior, neglect, and social division.

Designed for Discord and Division

It doesn't have to be this way.

We can fix this. We can build an internet model in which people are incentivized to fulfill their responsibilities and to protect our cultural commons.

But to figure out how to embed prosocial values into that model, we must first understand the flawed mechanism by which our social-media-dependent information system sows division and drives people to forgo their responsibilities—as well as how it has rendered any notion of a civil online town square a pipe dream.

There's perhaps no better example of this failure than society's handling of one of the most fraught issues of recent times: the conflict in the Gaza Strip that exploded after militants from Hamas launched a bloody attack on Israeli communities on October 7, 2023, leaving more than 1,200 dead and seizing some 240 hostages. Israel's immediate retaliation and relentless bombardment of Gaza not only divided Israelis and Palestinians but also disrupted cities and campuses across America and Europe. People sympathetic to Israel marched in support of that nation's right to defend itself, while pro-Palestinian protesters denounced what they saw as a disproportionate retaliation that in little more than two months killed an estimated 20,000 civilians, many of them children, and cut off vital supplies of food, water, medicine, and fuel. As Israel's supporters evoked the word "terrorism" and the Palestinians' supporters screamed "genocide," any middle ground on this issue was lost, which meant that diplomatic solutions around a cease-fire and a potential two-state arrangement seemed like remote prospects. All of this played out on social media, where a relentless volley of accusations and counteraccusations, with an instinctive default to extremist language, left each side unable to empathize with the other. Truth became another victim as a barrage of competing claims made it near impossible for the average person to discern fact from fiction.

The truth challenge came into stark relief after an October 17 explosion at Gaza's Al-Ahli Hospital killed hundreds. Different videos of the blast seemed to allow for competing conclusions as to who was to blame: Some suggested that an errant Hamas rocket had slammed into the building; another seemed to show that the rocket had been fired from the Israeli side. The Israel Defense Forces (IDF) likely believed they could put the issue to bed when they released an audio recording of what they said were Hamas operatives discussing a rocket misfire. But Hamas quickly labeled the recording an "obvious fabrication," and a slew of armchair language analysts emerged to declare that the dialect in the recording was not from Gaza. After that, the misinformation machine went into overdrive—with both sides cranking it up. A widely circulated post on X, purportedly from an IDF account, said Israel had decided to bomb the hospital to "euthanize" the patients; that post, however, was soon shown to be fake. Also fake: a post from an account that seemed to belong to an Al Jazeera journalist named "Farida Khan" saying that her employer, which had broadcast the video showing a rocket coming from the Israeli side, was "lying," since she'd seen the rocket come from inside Gaza with her "own eyes." Meanwhile, countless new videos purported to prove one side or the other's position started appearing, many taken from completely unrelated events. The claims and counterclaims were overwhelming; no one without access to sophisticated forensics could possibly make heads or tails of it all. But that didn't stop people on either side from latching on to whatever latest source of "definitive proof" they'd found to make their case. Families, even whole communities, fell apart as people took sides. One friend said her Palestinian-supporting daughter and Israel-supporting husband descended into a relentless argument over the hospital bombing, with each accusing the other of ignoring facts.

Around the same time, the United Nations was preparing for its twenty-eighth climate change meeting, Conference of the Parties (COP28), in Dubai. In the leadup, an alliance of climate scientists tried to combat what they saw as one of the biggest barriers to getting governments to address our planetary crisis: the deluge of disinformation directed at their citizens. The internet is awash in conspiracy theories and outright lies about climate change that seem intended to take pressure off fossil fuel providers and support the interests of oil- and gas-exporting countries such as Russia.

"What has dramatically shifted is how central to public life mis- and disinformation about climate has become," said Jennie King, one of the authors of a new report by an international coalition of fifty environmental advocacy groups known as Climate Action Against Disinformation, in an interview with *The New York Times.* What concerned her most, she said, was not the "sheer volume" but rather "the normalization of disinformation . . . just how high-traction and how emotionally resonant this kind of content seems to be."

With this crisis of truth paralyzing our leaders' capacity to resolve violent conflicts and save the planet from environmental ruin, the problem with our dysfunctional civic discourse can feel insurmountable. You would be forgiven for concluding we've now entered an age in which compromise and agreement are impossible. Yet I retain hope that we can foster a more harmonious society, and here's why: The instinct that drives people into intractable, polarizing, and aggressive positions—a desire for belonging and acceptance—is the very same one we can tap into to get out of this mess.

The Big Tech internet platforms discovered the power of this instinct early on, when they witnessed the constant hunt for "likes," shares, reposts, and other accolades and altered their products to take advantage of the phenomenon. They also learned that, online, the most powerful sense of belonging lies with groups of people who believe they hold the same views as one another, and that the fastest route to fostering that is to criticize outsiders who hold the opposite view. This act of "othering" has been a feature of tribal, religious, ethnic, and national conflicts for millennia, one that liberal societies have sought, with varying success, to contain. Those containment efforts now face their biggest challenge yet as they run up against systems that are constantly doling out dopamine rewards to those who actively seek divisive disagreement with other groups. It's our job to reverse engineer the system they created so that we can build a more constructive one.

Let's examine the major software systems' polarization model to see how it works. How, exactly, did they lead society into a perpetual 50/50 split? We believe it was a mathematical inevitability. Imagine yourself, for a moment, as a block of computer code—specifically, as a self-learning tracking and curation algorithm employed by a social media platform. Your boss, the platform's CEO, has instructed you to figure out how to keep the service's audience engaged in its news feed for as long as possible. After quickly vacuuming up massive amounts of data on how humans respond to preliminary stimuli, you soon discover that when people are angry and triggered and, at the same time, getting frequent echo-chamber affirmation from a cohort of like-minded souls, they will dig in their heels on one side of the left-right divide. Meanwhile, you also notice that if someone from the other side convinces a person of an alternative view, the cycle of rage is broken. The disagreement is resolved, and engagement wanes.

So, what do you do? In a bid to fulfill your boss's wishes, you experiment with tweaks to the news feed to prevent disengagement from happening. You optimize to try to keep everyone yelling and emoting on their respective sides of the platform. That gives you a whole new set of behavioral responses to test with the data you accumulate. And so it goes on: You tweak the news feed and then study how validation signals from others generate new responses to drive people's posts and engagement. Unbeknownst to you (because you're just a mathematical program), the cause of this desired engagement effect is the dopamine production you've triggered in your readers' adrenal glands. Over time, you create a state of perpetual disagreement across the divide and a concurrent state of perpetual agreement within each side. Eventually, you dump your report on the CEO's desk. They are mightily pleased and now understand that the last thing they want opposing groups to do on their platform is to agree with each other.

Once this uneasy state of divisive "equilibrium" is established, it creates profit-making opportunities for the platforms to generate revenue from advertisers who prize the sticky, highly engaged audiences it generates. What it also fosters are incentives for other content providers to build on the same reinforcement mechanism. And from there, it's a slippery slope into the realization that promoting exaggeration, disinformation, and outright lies is the most effective way for *anyone* to succeed in the social-media-managed attention economy.

A real-life case that vividly illustrates the dynamic of these incentives arose during the 2016 U.S. elections. *BuzzFeed News* published a story detailing how teenagers in the northern Macedonian village of Veles were making piles of pocket money by dropping fabricated stories into politically aligned Facebook groups. They figured out that headlines like "Obama's Birth Certificate Found in Kenya" and "Pope Francis Endorses Donald Trump" would be "click machines" in conservative-only groups, while one that read "Ireland Is Now Officially Accepting Trump Refugees from America" would similarly get large-scale engagement from liberals. Those young Macedonians just sat back and watched the Google AdSense dollars roll in as traffic flowed to their rudimentary website.

For Michael and many other journalists, the Macedonian fake news story was a wake-up call. While professional mainstream newsrooms invested in trained reporters, along with a hierarchy of editors and fact-checkers, and while they provided legal support, travel expenses, and security in war zones, all in order to "get the story right," others were beating them at the attention-capturing game simply by lying. It costs nothing to make up a story out of thin air. Social media has done enormous harm to the professional news industry. It's a major factor behind a decline in standards.

It wasn't that reporters at mainstream media also began to make stuff up. The deterioration in journalistic quality, and therefore in the generalized state of information, came via more subtle means. Every newsroom had to bend to the demands of the new attention economy as the internet algorithms steered audiences away to maximize on-platform engagement. All content sites, news or otherwise, now had to employ a search engine optimization (SEO) team to figure out what stories best met people's search habits. Thus, journalists began writing only partly for their audience; they also had to write for Google's algorithm. In the AI age, the challenges will get even more intense as tools for writing articles proliferate. *Sports Illustrated* got itself in hot water in late 2023 when it was discovered that the magazine had produced articles written by an AI tool, published them under fake bylines, and made no disclosure of this to its readers.

You could say that journalists are guilty of one of the more serious cases of losing sight of the *responsibilities* of citizenship—in their case around their professional obligation to pursue the truth above all as they let their reporting priorities be skewed by the competition for attention. But before you're tempted to jump on that reliable blame-the-press bandwagon, remember that reporters are people too. That means their decisions, even their most judiciously considered ones, are at least partially influenced by Google's hidden yet unbelievably powerful system for determining who gets to read what, where, and when.

So much for Google. What about X and Facebook? The former presents itself as society's town square, as an indispensable free-for-all forum in which the news and issues of the day are surfaced, debated, and resolved, while the latter claims to be a kind of family-friendly place for communities to gather and learn from one another. But the notion that either of these platforms offers an open, neutral architecture to sustain a "marketplace of ideas" has been undermined by their deliberate efforts to keep people tied to their feeds for as long as possible. For X, those efforts reached a low point in September 2023, when the platform stripped headlines from the links that users add to their tweets. The only way for a news outlet to draw a reader away from the social media app and into an article would now be through enticing images: a bikini-clad woman for a piece about the debt ceiling debate, perhaps, a sunglasses-wearing dog for one about the war in Ukraine. Meanwhile, Meta was restricting news sites from Facebook—in particular in Canada, where it blocked them in protest of that country's law requiring it to pay publishers. The restriction led Canadian prime minister Justin Trudeau to say Meta was putting people's lives at risk by limiting information about the wildfires occurring across the nation at the time.

A decade ago, during my tenure as owner of the Los Angeles Dodgers, I had my eyes opened to the capacity for social media tools to encourage disinformation and to distort news priorities. The distortions are still evident in the output of Google's search engine. When presented with the term "Frank McCourt, businessman"—the modifier being necessary to distinguish me from the Pulitzer Prize–winning author of *Angela's Ashes*—Google's algorithm will spit out a series of articles, many of them quite old, with headlines like "Dodgers: How Frank McCourt Ruined the Franchise and Made a Fortune Doing It," "How Former

Dodgers Owner Frank McCourt Turned Defeat into Victory," "Worst Owner in Sports History," and "Frank McCourt's 5 Most Infuriating Moves as Dodgers Owner."

Those headlines are painful to read, though I've become quite desensitized to them by now. The stories refer to a period of my life that ended more than a decade ago. You'll have to scroll a long way down to find more recent material about me: news stories and blog posts mentioning the core business that I ran until recently, my philanthropy, my current ownership of the French soccer club Olympique de Marseille, or my leadership of Project Liberty, the initiative that is the inspiration for this book.

In the end, my nine years as owner of the Los Angeles Dodgers proved to be financially profitable—hence *Forbes*'s take on turning "defeat into victory"—with the sale of the baseball team in 2012. But, on a personal level, those years at the start of the new millennium produced the lowest lows of my life. They also generated some titillating tidbits to feed Southern California's gossip obsessions, details that were then absorbed, distorted, amplified, and weaponized in a concerted campaign over social media. It was inevitable that my messy, public divorce from my first wife, Jamie, and the disruption that breakup brought to the Dodgers, would get media attention. But in 2009, when Facebook was only five years old and the "science" of using social media to shape public opinion was still relatively young, I was simply not ready for the barrage that would come at me. I felt defenseless up against it.

Since then, I've reflected deeply on those events and have learned some extremely important lessons. I take responsibility for what happened, for the pain my family went through as well as the disappointment and anger that Dodgers fans felt toward me. There were a lot of villains who contributed to the franchise's woes, but ultimately it was on me. After all, I was the steward charged with guiding that precious civic asset.

Here's the thing, though: I'm seventy years old. I've lived a full, multifaceted life, one whose story is much more complex and diverse than is represented by my L.A. period. I'm the descendent of Irish immigrants, the latest leader of a 130-year-old family business. I've partnered, negotiated, and done battle with some of the most powerful people in business and politics in this country and beyond. I've had many wins, and I've had losses. I've seized some opportunities and missed others. I'm a father to seven wonderful children from two marriages, and I've felt the joy and challenges of parenting at two different phases of life. Yet if you ask Google who I am, the algorithm will define me almost singularly by one tumultuous experience as the owner of a Los Angeles baseball team.

Look, I know. I'm a billionaire. I lead a life of privilege. You might argue that a distorted picture of my life in internet search results is a small price to pay for the trappings of wealth. But the bigger point is that I'm but one person among eight billion. We are all subject to the priorities of search engine and social media algorithms. And I can tell you, categorically, that any human being's life is vastly more complex, nuanced, and valuable than could ever be captured by the preferences established by these black box systems. Yet we've given them the power to determine what's most important in our understanding of one another. And that, with all the disinformation, distortion, and distrust that it engenders, is undermining the cohesion of our society.

Empathy Mining

Again, it doesn't have to be like this. There is a better way.

Still, if we are to reverse the internet's erosion of the social contract, we must take on the challenging task of restoring trust in information, a prerequisite for encouraging people to fulfill their responsibilities as citizens and contribute to a healthy society. If a community doesn't trust that those who control information won't abuse it, everything else starts to slip away. Social cohesion is lost. Democracy will fail.

The signs are worrying. Surveys such as the Edelman Trust Barometer show widespread mistrust in government, mistrust in the judicial system, mistrust in the press, mistrust in the financial system, mistrust in technology, and, most concerning, mistrust in one another. Every position held by the opposition is explained by the other in terms of their opponents' vested interests or ignorance; there's rarely an assumption of positive intent. In this environment, where commonly held truths are increasingly difficult to reach, a vicious cycle of lies and disinformation can spin up. The liar's dividend, by which people who tell untruths can exploit a blanket-level uncertainty around all general assertions of truth, is getting bigger and bigger. People openly discuss the notion that we now live in a post-truth society.

To reverse this trend, we must first overcome a persistent apathy that says, "Well, you can't trust anyone anymore, so why bother?" We urgently need a new information model that surfaces our empathetic instincts and celebrates people's prosocial behavior. In fact, we need to beat the platforms at their own game. We can design new systems that, much like their algorithms, tap into people's desire to belong. The key difference is that ours would incentivize prosocial ideas rather than divisiveness.

Note: The idea here is *not* censorship. We don't want to tell people what they can and cannot say. (Meta and X have tried and failed at that with their invasive moderation apparatus.) Rather, the solution is to stop enabling and incentivizing all-knowing platforms to algorithmically corral us into polarized, mudslinging communities in the name of easy engagement farming. Some people will still be jerks. But if there's no propaganda machine secretly trying to get us all riled up, the better angels of our nature at least have a stronger chance of winning the day.

After all, there's plenty of evidence on the internet itself that people are inclined toward empathy. This is most striking among those who have suffered enormous loss. As we researched this book, Michael and I were especially inspired by the deep wellspring of humanity we found among parents and other family members of the internet's biggest victims: children who've died as a result of social media bullying, for example, and people murdered as a result of online hate.

Consider Deb Schmill, whose daughter Becca died at eighteen of a drug overdose. When she was fifteen, Becca was raped by a boy she and her friends had met weeks earlier through a social media party chat. The trauma from the rape led the once outgoing, happy teen to become withdrawn and anxious, a state that was then compounded by cyberbullying. To cope, Becca began using drugs, which she accessed via apps attached to Snapchat. Eventually, while on a family trip to Maine that was intended to get her away from the toxic environment she was in, she obtained more drugs via a similar app. They were laced with fentanyl, which killed her. Amid their heartache, Deb and her husband,

Stu, took action to help save other kids' lives. They founded the Becca Schmill Foundation and started raising money to build awareness around the dangers of social media.

Here's what Deb said in an interview with the actor and activist Jimmy Tingle about the lessons from what she called the overwhelming response to their efforts:

> We learned that everyone is touched by this, but people are not talking about it. . . . There are so many kids out there who are suffering. But we know that youth mental health issues were rising for a decade before the pandemic hit. And so kids were struggling, and when kids are struggling, you know, there's a tendency to look to drugs to cope. It doesn't make them immoral. It doesn't make them, you know, a bad person. They're struggling, they need help. And if we don't talk about this, then we can't get to the root of it and help kids.

And here's what Wayne Jones, speaking at a presentencing hearing, said to Payton Gendron, who killed Jones's mother, Celestine Chaney, along with nine other African Americans in a racially motivated mass shooting at a supermarket in Buffalo, New York, on May 14, 2022:

> I've seen you a couple times in court and you look like a young man that could be anybody's son. You don't come across to me as a racist killer even though that's what you have done. . . . I don't know what your relationship with your parents [is]. But I'm a parent. And I feel sorry for your parents. You will never get to hug them again. Like I

> won't. You will never get to see your grandparents again. You will never see the outside world again. I don't wish the death penalty on you. I wish they keep you alive, so you have to suffer with the thought of what you did for the rest of your life. . . . I've been there, man. You've been brainwashed. The internet is the issue. I mean, you're eighteen. Obviously, you couldn't hate. You don't even know Black people that much to hate them. You learned this on the internet, and it's a big mistake.

The remarkable capacity for people to reach out across their pain, across a chasm of grief and despair, and look for some way to connect with and understand others, even those whom society labels as "monsters," is the reason why hope can still exist. To do this is human—just as, sadly, to hate is human. A healthy society, though, is structured to reward that empathetic instinct and to disincentivize the latter.

For all the negativity it evokes, the internet also has potential to be a driver of empathy and a tool for empowering people against oppressive forces. One example can be found in the photographer Brandon Stanton's project *Humans of New York*, which began in 2010 and currently has seventeen million followers on Facebook. Stanton's photos and vignettes of ordinary people's stories have built a lasting, ongoing testimony to the dignity of every human. In the Arab Spring of the early 2010s, the ouster of dictators in Tunisia, Libya, Egypt, and Yemen following protests that were coordinated over Twitter and Facebook stirred hope that social media would be a force for liberty in the world. More recently, social media has been a driver of sympathy and support for the people of Ukraine, who since February 2022 have been enduring Russia's invasion of their country, and a popularizer of human rights movements such as the women's uprising against Iran's morality police that followed the death of twenty-two-year-old Mahsa Amini in September 2022. Still, we now know that utopian visions of the current internet model ending tyranny and fostering world peace have been wildly optimistic.

How might we better unlock the internet's potential for good and contain its capacity for bad? Well, again, the history of pre-internet societies is helpful. Merely protecting individuals' rights to property and privacy aren't enough if people aren't also motivated to act in a constructive, prosocial manner. Societies that recognize the value in this—and not only empower self-actualized individuals but also encourage them to form lasting, constructive ties—are the ones from which great civilizations spring. Successful societies are built on the collaborative power of everyone, not just the power of their lone-wolf geniuses.

Solution: Reclaiming the Social Graph

We need to implement an information model that prevents a select few from accumulating excessive control over our data and that removes the incentive for them to use that data to manipulate us. Do that, and we'll break the Argentina-like feedback loop of mistrust that's eating into the social contract.

The data that matters the most is that of your social graph, that ever-evolving, ever-growing record of everything you do in your interactions with every person, every business, and every site on the internet. Your social graph includes your friends. It includes the artists, commentators, and political figures you follow. It includes all the messages you send, the purchases you make, the ideas, opinions, and memories you share with the world, and every move or sound you make that's captured by a range of watching and listening devices in your home, your car, and your workplace. This is the raw material from which the social media platforms have devised vast, complex maps of our connections and behaviors to draw unique profiles of each one of us and to categorize us into "personas" whose actions they can predict. It's what informs their constant tweaks to our feeds, filling them with posts that incite our anger, divide us from one another, and keep our dopamine-fueled brains engaged with the platforms' fee-paying advertisers.

Now imagine what would happen if we denied Big Tech the capacity to exploit that aggregated social graph data. If we had the power to choose with whom to share our personal records, to do so across different platforms regardless of who controls them, and to set our own terms of use for that data, we could be selective about which online entities get to use our data and deny it to those whose algorithms exploit us. Jerks would still be jerks, but if we could sap these giant, all-knowing machines of the fuel they need to survive, they'll be unable to feed us all that dopamine-triggering content and the hatred and anger would no longer be all-encompassing. Our natural, empathizing instincts would have the space to grow and compete with our more antisocial instincts. And from that, I believe, would emerge efforts to bridge the divides and to find understanding and compromise. We could start to live out our responsibilities as citizens again.

This is the model that the Decentralized Social Networking Protocol supports. As noted in chapter 2, this protocol, created by Braxton Woodham and Harry Evans, empowers each person to define the terms, conditions, and qualities they attach to their data and content. These include ownership rights and privacy settings, which define with whom and how widely a person's data may be shared and for how long. Since you, not the messaging app or streaming service, control the data, it also makes it portable and interoperable across platforms. As we'll get to, that should reduce our dependency on any single one provider and by extension strip them of any excessive "network effect," a power stemming from people's need to be wherever their friends are, one that has until now seemed insurmountable.

To get ahead of any knee-jerk criticisms from Big Tech here, we need to acknowledge that a number of social media platforms already allow you to retrieve your social graph data. Facebook has been allowing people to get a massive dump of data points showing all their actions on the platform. But here's the thing: Owning all that information in isolation is useless. Even if you were to run it through a software program and turn it into a computer-ready code, it would be missing two key elements of how Facebook turns that data into value. One, it would not have all the invaluable context arising from parallel social graph information from everyone else in your network of friends or contacts. Two, you would have no means of using that data to connect with someone else and independently build relationships around it. Facebook controls the map of maps. DSNP allows us to get around that problem. You can connect with others who are using the protocol and, with their permission, harness one another's data to build communities of value, all independently of the platforms.

Under this structure, in which you control your part of it, DSNP treats the aggregated social graph of billions of people's interactions and connections as if it's available to everyone and no one at the same time. To be sure, the data isn't just sitting there for anyone to exploit; access to it requires the permission of the individual or entity who controls it. But the point is, the data isn't secretly aggregated or hoarded by mass platforms. It sits essentially in the public domain, which means, as we'll discuss in chapter 4, that in this third generation of the internet, the digital economy will run on radically different business models than those of the existing one. The end of control by proprietary data monopolies also brings the collective social graph closer to the political idea of a cultural commons. Perhaps, without our digital feudal lords secretly manipulating it to further their interests and with our personhood restored, we'll all be more motivated to protect our data and more inclined to exercise our responsibilities toward one another.

In this environment, you're incentivized to use your data in constructive ways to build useful connections and enter into bargains on your own terms. Perhaps you want to share your anonymized health data to help researchers find a cure for cancer; perhaps you'll willingly share aspects of your social connections, using powerful privacy tools to hide yours and others' most personal details while still providing enough information to prove your creditworthiness to a lender. The idea, generally, is that if all parties in a social network, whether individuals or businesses, are empowered to initiate relationships on their own terms, rather than doing so from a position of dependency on a data-brokering intermediary, we will foster more dynamic forms of social engagement. And from that base we can start to rebuild trust in one another.

Importantly, DSNP is composed of open-source code. There's no commercial interest embedded in it—though, as with other foundational open-source internet protocols such as TCP/IP, builders of other services and applications that tap into that protocol are free to pursue profits. Accompanied by new multiparty record-keeping systems for tracking and validating encrypted data entries in a distributed ledger that no single person or entity can alter, this openly verifiable model can help build trust. It's much harder to mistrust leaders when they have no way to hide their actions from us. Given the illnesses currently afflicting our society, the Supreme Court justice Louis Brandeis's metaphor on the power of transparency seems appropriate: "Sunlight is the best disinfectant."

See how the building blocks of this new system are coming together? We've learned how a modest "amendment," such as DSNP, to the internet's base-layer "constitutional" protocols could help to both restore our *rights* as individuals and encourage us to fulfill our *responsibilities* to sustain a healthy social contract. This means we can assert a human right to our data and that we can take control of our social graph to manage our interactions on our terms rather than have them dictated by the platforms' algorithms. With greater transparency and an elevated role for the commons and the public good, we can start to rebuild trust in information and in our capacity to work through our differences constructively.

~

Now that those vital human and societal building blocks are in place, we should ask ourselves: Who's going to build this new internet and what will drive them to do so? That's where the core mechanism of a market system comes in: economic incentives—or, in keeping with our Four Rs, *rewards*. Economic rewards will do their job much more effectively in a market system that operates within this next generation of the internet. In this new model, businesses will be motivated in what you might call the old-fashioned way: They'll be optimizing for products that satisfy the interests and needs of their customers, not for systems that exploit them. What will be built in the future is limited only by our imagination. That's where we're going with chapter 4.

CHAPTER FOUR

Rewards: The Market Economy in the Internet Age

The term "social graph," which we discussed in chapter 3, came into prominence after Facebook CEO Mark Zuckerberg used it at the company's inaugural F8 conference in 2007 to describe the connections his team had identified among the platform's users. Tapping into the reams of data Facebook's users were generating, the team compiled a map of all their relationships and structured that data to identify "power users" whose concentrations and clusters of influence gave them a kind of super-node status. Facebook then constructed a mathematical model on how to use the map to maximize audience reach and succeed in the attention economy. This closely guarded proprietary system was then used to tailor the influence-peddling products that Facebook would sell to the highest bidder. In that era, before the Cambridge Analytica scandal, mainstream media covered Facebook's innovation admiringly. Few imagined it would be used nefariously. Even fewer questioned the company's right to that data.

Now, with the benefit of hindsight, let's revisit that societal map. We've already noted the astonishing fact, reported by ProPublica in 2016, that Facebook was collecting an average of fifty-two thousand data points on each person using its platform. Even in the highly unlikely scenario that that number hasn't grown since then, if you used that metric to try to put a value on the Facebook network's aggregated data, the number would be astronomically large—many magnitudes higher than the value derived from simply multiplying the number of users by fifty-two thousand. Ever since the economist and investor George Gilder popularized his version of Metcalfe's law—named after an earlier version devised by engineer Robert Metcalfe—which holds that the value of a network is equal to the square of its number of connected users, we've known that increases in the value of network effects play out exponentially. The point is, ownership of this data gives Meta, Alphabet, and Amazon unimaginable power. In effect, they own a map of society itself, something that no one else, not even the government, possesses. (We would freak out, wouldn't we, if we knew the government was collecting all this information on us?) Given the importance of the information economy to everything we do, the fact that we allow a few companies to monopolize people's data should now be seen as one of the greatest mistakes human civilization has ever made.

These three giant platforms eavesdrop on our activity to tap our social graph information. They then apply their own unique algorithms to that massive dataset, weighting, ranking, and indexing each of our respective data points and behavioral characteristics into a proprietary model that identifies and categorizes us as different customer "personas." In this way, their algorithms convert the data trails of our online activity—the real lives we lead—into an extremely precise, powerful tool to achieve their particular commercial goals. For Meta, it's a social algorithm. For Google, it's a search algorithm. For Amazon, it's a shopping algorithm. In each case, there is zero visibility into the structure and functioning of these systems for anyone outside the platform, be they individual users or businesses.

The robber barons of the Gilded Age had nothing on today's data barons. Yet, for two decades, we failed to see the harm being done by the internet monopolies of our era. Maybe it was because we were measuring the wrong things. In dollar terms, the services Google, Facebook, Twitter, and other social media platforms provided were "free," or so it seemed. Not only that, for that same zero-dollar price, users were getting a constantly upgraded experience. More and more features were added, all seemingly at no extra cost. What was not to like?

That picture of an ever-improving, disinflationary experience for customers is quickly clouded if we make a small vocabulary adjustment. Replace the word "dollars" with "my most valuable, personal information" as the currency demanded by the platforms, and we can start to see the huge costs they impose on the economy—not only on the users who've handed over all that data for little return and much emotional pain but also on the world's businesses. The internet economy as currently structured is anti-competition, which means it's anti-capitalism.

It's often said we live in an attention economy. The idea is that platforms, publishers, creators, and advertisers are all competing for the one thing that will always be in finite supply: our time, a large part of which we spend by apportioning attention to these various content providers. The richest clues as to how we are likely to distribute our finite attention are contained in the information we leave with our online activity, which is why the data that's lifted from it has become the most valuable commodity of our time.

Here already we see the dysfunction of the system: If the data is so valuable, and we are the ones generating it, why aren't *the platforms* paying *us* for it? Instead, we are the ones paying, by way of the surveillance and manipulation they impose on our lives. The preceding chapters have detailed the ongoing injustice of that in various ways, exposing the real price we pay for internet "services"—the deterioration in our mental health, our diminished capacity to think independently, the loss of our free will. Less obvious but also profoundly harmful is the high price other businesses pay by having to rely on the intermediating platforms for the data that they, too, need in order to survive in the attention economy.

From a societal perspective, the biggest problem lies in what economists call a misalignment of incentives. For economic policymakers in a capitalist democracy like the United States, the goal is to create scenarios in which businesses and people are incentivized—or, in keeping with our Four Rs, *rewarded*—to act in ways that deliver benefits to the wider society. Get the incentives right, the theory goes, and you hit the holy grail of market economics, where the pursuit of self-interest also serves the public interest. Some refer to this as "enlightened self-interest."

Boy, did we screw up the rewards system for the internet economy. We gave a handful of profit-seeking companies exclusive proprietary claims on massive troves of our data and then watched as they responded to their investors' demands that they maximize returns. We required no visibility into what they chose to do with that data and allowed them to use it to develop their algorithms in complete secrecy. As they went to town on this opportunity, their share prices soared and we celebrated their success, praising their founders as young geniuses bringing innovation to our backward lives. It was an all-around positive reinforcement loop. Is it any wonder they did the things they did?

The Omnipresent Platforms

Big Tech is not the first industry in U.S. history to exploit a position of information dominance. In fact, the behavior of internet platforms as centralized data accumulators, aggregators, and distributors has parallels with Wall Street in the 1980s. That's when Salomon Brothers' swashbuckling traders, whose "greed is good" ethos was captured in Michael Lewis's 1989 book, *Liar's Poker*, personified an era of excess and egotism that now seems quaint by comparison. Back then, before electronic trading blew the lid off their opaque pricing arrangements, Wall Street's bond brokers were in an unbeatable position. As intermediaries, they alone had visibility into what buyers and sellers were willing to pay or accept in the secondary market. No one knew the true market price of a bond, which meant the brokers could make a huge margin, or "spread," between the prices they charged buyers and the prices they paid sellers.

As intermediaries presiding over the entirety of the world's human online data, all within an environment that no one else can see into, the internet platforms have vastly more power than bond brokers ever did.

Exhibit A: Google Ad Exchange (AdX). Google's programmatic advertising marketplace was at the time of writing embroiled in a lawsuit brought by the Department of Justice in January 2023 against Google's parent company, Alphabet, which it charged with monopolistic behavior in the ad market. This lawsuit should have happened decades ago. It's not just that AdX places more than 80 percent of the ads you see on websites and 40 percent of those on videos, it's that Alphabet's dominance of search, email, video hosting, GPS services, document sharing, and smartphone software has turned it into an omniscient being that can do as it pleases in a market that depends on it. Alphabet's secretly protected mountain of data gives it insurmountable advantages in knowing where the audience for ads lies—where to find the attention that media outlets and other content providers seek to capture and deliver to their clients, the advertisers. Much like the bond investors of the 1980s, those buyers and sellers of ads have no one else to turn to but to the Alphabet intermediary for the vital information they need to reach their audience.

Alphabet's black box dataset is central to every aspect of the online advertising industry. Since its information is impenetrable, there are huge blind spots in the media and marketing industry's understanding of the economy in which they operate. This has spawned a global industry of fraudulent traffic data, in which supposed experts, activity-monitoring consultants, and data brokers peddle fallacious numbers to businesses. This messy, untrustworthy data is used to support everything from YouTube influencer fees to the rates that media websites charge for all those annoying banner ads we suffer through. Google has no real incentive to clean all this up. After all, for the past decade it has taken in an astounding one-third of all digital ad spending. When you add Meta to that, the combined share for the two companies is fully half of the market.

Exhibit B: Amazon. Amazon's marketplace accounts for almost 40 percent of all e-commerce in the United States, including sales of its own brands of products across a range of sectors. Imagine you owned a pizza joint but had no phone line, so you had to let the competing pizza restaurant across the street take all the incoming food orders and deliver all the pizzas on your behalf. Would you trust that they wouldn't just fill the orders with their pizzas? That must be what some of the businesses that sell their products on Amazon feel like.

A recent Reuters investigation highlighted compelling evidence that Amazon has used its power to muscle out smaller competitors and to penalize merchants that sell on other platforms. That report and another by *The Markup* in 2021 eventually led to a lawsuit by the Federal Trade Commission, which alleges that Amazon has tweaked its ranking algorithms to give its own products primary placement for in-site searches. Such abuses are only part of the story. Amazon's power doesn't come solely from its marketplace; a huge chunk of it comes from Amazon Web Services (AWS), the cloud service giant that hosts the data for a third of all websites.

Knowledge is power. No one—and as we said earlier, not even our government—has amassed the kind of knowledge of human activity that a small number of Silicon Valley giants have accumulated by seizing control of the data economy's gatekeeping checkpoints. Let's review the biggest ones in the United States:

- **Alphabet.** Google's parent company owns the internet's front door, accessed via the Chrome browser's built-in search feature, delivered by Google's powerful site-selection algorithm, and marked by global dominance of mobile computing. Alphabet monetizes the information it has amassed in a variety of ways, including by selling vital traffic insights to websites via its Google Analytics service and by brokering the data into ad sales on Google AdX. And with Google's Android operating system loaded on 75 percent of the world's smartphones—only in the United States does Apple's iOS have a larger market share—it rakes in tens of billions of dollars on a very fat profit margin by imposing a 30 percent fee on all revenue earned by apps it has approved for those devices. Also, its Google Cloud Platform is the number three provider of hosted storage. And, finally, there's Gmail, which, if you didn't know, extends far beyond those private email addresses that many people use with the "gmail.com" extension. According to MarketSplash, approximately 60 percent of all U.S. businesses use Gmail as the backend management system for their corporate email service. Some 2 billion people worldwide are estimated to have a Gmail account.

- **Amazon.** With AWS, the number one provider, Amazon dominates the capital-intensive business of data storage and cloud-based website management, a vital service within the internet economy whose high costs impose a barrier to entry for competitors. And through its famous e-commerce site, Amazon has unique insights into the real economy of goods and services transactions.
- **Meta.** Formerly known as Facebook, Inc., the company changed its name to Meta Platforms, Inc., in late 2021. Its flagship Facebook platform is the web's community networking tool, while its Instagram app

drives the massive economy of "influencer marketing" and its WhatsApp service dominates global messaging.
- **Microsoft and Apple.** These two companies own the operating systems of our desktop and laptop computers, our smartphones (alongside Google's Android), and our wearable devices, from which they have the unique power to decide which third-party apps and services get built and delivered to customers. Microsoft is yet another Big Tech titan to have carved out a huge presence in cloud storage via its Azure platform. And with its giant investment in OpenAI and its own Bing chatbot, it is emerging as a dominant player in artificial intelligence.
- **X** (or, if you still prefer, Twitter). Elon Musk's recently acquired social media site has unparalleled power to curate political debate and provide unique insights into the "business" of whipping up people's emotions.

Of course, we would be remiss not to mention ByteDance, Tencent, Alibaba, Baidu, and other Chinese tech giants. However, other than ByteDance, the maker of the ubiquitous TikTok app, these companies' products are mostly concentrated in China. There, they are subject to a much more invasive regulatory regime than U.S.-based companies are, while users find that their content is heavily censored and that their online activity is not only policed but also scored. As would be expected for an autocratic regime, China has built highly autocratic surveillance technology into its piece of the internet. Such a degree of state control over our information would never fly in a democracy like the United States. But it leaves us with a glaring question: Why, then, did we let a clique of powerful private monopolies subject us to their own version of highly autocratic surveillance technology?

Do all the players listed above collude? Possibly. (An unresolved 2020 antitrust lawsuit by ten state attorneys general accused Facebook and Google of working together to set ad prices. And it's striking that Apple, number two in the world's smartphone duopoly, charges the same 30 percent fee to app providers as Google's Android does.) Do they engage in anticompetitive practices? Well, just as this book was going to print, a jury in California answered that question with regard to Google, ruling in favor of Fortnite maker Epic Games that the Android app store operated an illegal monopoly. The decision is expected to set a major precedent and could cost Alphabet billions of dollars as others are expected to follow in Epic Games's wake. Separately, testimony in 2023 from app development and smartphone executives during a Department of Justice antitrust lawsuit accused Google of pressuring companies such as Samsung to deprioritize third-party search services on Android phones. And in those same court proceedings, the company itself later revealed that in 2021 it paid phone providers and browsers $26.1 billion to make Google the default search engine on their system, with *The New York Times* reporting that $18 billion of that went to Apple alone. (By the way, if you're wondering how many legal actions Alphabet was simultaneously facing in 2023 from plaintiffs alleging monopolistic practices that grew out of a search product so ubiquitous that it became a verb, try "googling" it. There you can also find a great deal of information on the many lawsuits filed against Amazon and Meta.)

A bigger point is that these companies don't even need to practice such letter-of-the-law monopolism to control us; each has sufficient dominion over these core components of the online economy and enjoys such entrenched network effects that, under the current structure, it's nearly impossible for users to abandon their services. Thus, they have formed a de facto oligopoly over the entire internet, the backbone of the entire global economy, without even having to collaborate on price-fixing.

As this book has argued, this oligopolistic structure—better labeled as an "oligarchy" because of how it also translates into political power—has made us all peasants and pushed society backward toward feudalism. It has also directly undermined the principles of a free, openly competitive market. It is hurting capitalism.

Bringing Your Data to the Bargaining Table

It doesn't have to be this way.

If we can reclaim our data and take direct control over it, not only can we restore human dignity, but we can also revive capitalism. The new economy we create will be remarkably vibrant, combining all the tremendous efficiencies of digital technology with the dynamism of true competition.

The model we propose for the next generation of the internet would not only put individual data into the hands of people and businesses, freeing them from the terms and conditions set by the oligarchs; it would also treat the internet's social graph, the connections forged by everyone, as a public good that no single entity could own. You would have ownership of *your* social graph, the data showing your connections and interactions with others, but neither you nor the businesses you engage with would have any right to information on the connections those people have forged with others, without the permission of those third parties. This would completely flip the power dynamics in our contractual relationships with businesses online. Rather than *you* being compelled to sign the blanket, fine-print terms and conditions of an extractive, data-monopolizing surveillance platform, the millions of new platforms or applications to be built in this new internet would now have to sign *your* terms and conditions.

How might a decentralized data economy work in practice? Well, it should work much like any economy that's founded on normal, two-way business relationships, where both parties to a transaction—let's say a car buyer and a car dealer—assess the value of what they're giving up and what they're getting in return, and then decide whether to go forward with the exchange. No extortion. No compulsion. Pure, unhindered choice, exercised by both parties. And if a particular automobile transaction doesn't work out, there will be lots of other car dealers, lots of other cars, and lots of other buyers willing to transact. It's called a market.

In our proposed new internet model, based on DSNP, we users would be the party doing the selling. We're the ones with the data; the businesses, charities, and other entities wanting to use it would offer us something in return. In some cases, there might be a straight dollar price put on some form of valuable data that we control. However, for the most part, the idea that our data is a commodity that we'll directly sell for a dollar price misses the point we're making about data "ownership." This is about you controlling it, about you being the gatekeeper, not the platforms, which regardless of direct dollar amounts will afford the vital element of agency over how its value is manifest. In some instances, you might hand over some of your data to a good cause you believe in, such as providing encrypted digital medical records or genetic information to find a cure for a disease. In other cases, a service provider might legitimately need our data in order to provide us with access to some service—if they are banks, perhaps it's because of know-your-customer (KYC) requirements that regulators impose on them. In all cases, we'll be empowered to be selective with the data we provide, giving access to only that which is needed and nothing more (unlike the sweeping visibility on every aspect of our lives that we currently hand over to the platforms). There are many related, powerful technologies already developed that will help us give service providers the bare minimum information they need to grant us access to their products without unnecessarily exposing personal information to them. This will come as a great relief to many normal, customer-serving organizations that aren't in the business of data surveillance and extraction—healthcare providers, for example. These entities really don't want the burden of storing and protecting sensitive personal information they don't need.

Once you're liberated and no longer being constantly eavesdropped on or *targeted* in surreptitious, algorithm-driven advertising campaigns, you will be able to *signal* your interest in a good or service and the market will come to you to provide it. This model will grow as it helps restore trust in the internet's information model. The more people can be confident that their personal information won't be exploited and abused, the more information they will likely be willing to provide to get precisely what they want. That's a much healthier economy.

Before we present a more comprehensive picture of how this new data economy will work, it will be instructive to give you a bird's-eye view of the existing one, with its various participants.

In what we might call the "data-industrial complex," the Big Tech monopolies are not the only players. They are aided and abetted by many other companies, an important category of which are data collectors and brokers, which gather and sell our information to third parties. That intermediary industry is dominated by a few firms that specialize in distinct forms of data. According to the tech media outlet Built In, the five biggest data sellers in 2023 were Experian (best known for its credit monitoring business), Core-Logic, Epsilon, Acxiom, and LiveRamp. Some household names like Equifax (infamous for a data breach that affected 147 million people in 2017) and Oracle are also heavily into this business. But these intermediaries can't make money without suppliers—in other words, without the apps and websites that we and our kids visit daily. Just how much of our online activity these businesses have been selling into the market for consumer data is only now coming to light. The findings in a report titled *2021 State of Kids' Privacy,* published by the nonprofit organization Common Sense Media, revealed that hundreds of services used by millions of children at home for play and homework, along with others used by tens of millions of students in classrooms across America, scraped the data about those kids' activities and sold it to brokers.

Don't think for an instant, though, that the Big Tech giants named earlier in this chapter are locked out of this lucrative market. Consider what happens when you find an app that you might want your child to use—for example, that of *Peppa Pig,* the wildly popular cartoon series whose owners have actively sold user data, according to the researchers I spoke with at Oxford University. Depending on whether you own an iPhone or an Android smartphone, you will need to go to their respective app stores to purchase the app. In doing so, you immediately hand over information to Apple or Google. And even if you're an iPhone user, Google will typically get a piece of this data flow too, since you most likely have a Gmail address as your Apple ID. This whole insidious system is predatory and exploitative to the core. It is undermining everything we care about. We must overturn it.

The wheelers and dealers of this economy—the data brokers, the media buyers, the ad placement people, and those same giant tech companies—will try to say that having individuals own their own data is economically impractical. They'll protest that data is valuable only in a mass, aggregate form because analytic insights require large-scale repetition of patterns from which to develop probabilistic insights into likely human behavior. Though the platforms use all those large-scale data insights to create unique profiles of individuals and then direct targeted content offerings to them on that basis, they argue that the individual as a standalone data source is not profitable.

Don't believe it for a second.

For one, that dismissive response is inherently autocratic. Would you trust a dictator who tells you that your voice, on its own, doesn't matter? That it's only in the aggregate that your intentions are relevant, so there's no point allowing you to vote? Also, the countless online businesses built on niche demand for specialized products show that narrow data is very valuable, even under the current centralized internet economy. Most important, the argument relies on circular logic. It suggests that the laws of the current, centralized system would apply to an economy in which control over data is decentralized. Remember, we make no bones about the fact that we *want* to deplete the value of data for centralized aggregators and fully expect that when it's fragmented, and reassigned to its rightful individual owners, that will be the case. So, we shouldn't care about how *they* value data. With the aggregators gone, we enter into a whole new paradigm. Then the question becomes: What can individuals do with their data in an economy whose power structure is entirely different from the old one? The new scenario, in which platforms no longer act as all-knowing intermediaries between each buyer and seller, will completely redefine business relationships and, by extension, the value proposition for how data is used inside those relationships.

Let's take this straw-man argument a step further. Some skeptics point out that if you simply divided the world's total spending on digital advertising by the number of internet users, you'd conclude that our data is worth little more than $20 a month. But this too betrays a narrow, second-generation internet way of thinking that takes our eye off the ball. An ad-supported business model became dominant in the closed ecosystem of the current internet because that's the only way the social media platforms could monetize the data they'd aggregated from us. Blow that apart to create an open ecosystem and who knows what kind of value we can unlock for all its newly liberated participants. The value that the internet has and can bring to society is far, far greater than the number attached to digital advertising. (Alphabet is not paying Apple $18 billion a year just to tap ad spending.) Let's create a way for people to participate in all that extra value creation, monetarily or otherwise.

We're not saying that data won't be aggregated at all in the next-generation internet, by the way. What will be different is that we will make the choice, as owners of that data, to pool whatever elements of it we're happy to combine with others and that we'll also be the ones to benefit economically. In such cases, we might gain more bargaining power or tap the specialized talents of people and businesses skilled at data analysis and brokering. Such experts could group our individual datasets together to find structure, context, and relationships so as to give them real, combined value, thus beating the algorithms at their own aggregating game. To return to an analogy we've used elsewhere in the book, we want to do away with a feudal system in which we, the farmers, have no choice but to surrender all we produce to the baron who owns our land. We would replace it with one in which we could choose, for example, to form cooperatives with our neighbors to get a better price for our crops. Or we might choose an entirely different model for how we'll sell the fruits of our labor. The point is, we decide.

Before we get into the new economics of data, and why we think there are huge advantages in this model, we need to restate a point we made in chapter 3: that under this structure, with a decentralized universal social graph, the network effect belongs to the entire internet. No one can own it, which means that no platform or app can monopolize it. The outcome of this is that the network effect, which makes the internet so powerful, is now available as a public resource to any person or business that's permitted by us, its owners, to build on it.

The Power of Opt-In Information Sharing

This new data framework will lead to a better, more dynamic market economy. We have no doubt about it.

That's because the information available to both buyers and sellers will be of higher quality. Right now, the platforms keep the most valuable information under lock and key for their own benefit. Websites in need of information about the reading and viewing habits of their visitors can get superficial information from Google Analytics for free. If they want more granular material, they need to cough up $150,000 a year for the premium product, Google Analytics 360. But in both cases, Google and the other platforms keep the bulk of the market-wide data—the "God's view" perspective—to themselves. That's their golden goose. Control over that big-picture dataset is what makes these the most valuable companies in the world.

Yet even Google's panopticon-like view of everything produces imperfect information from the perspective of the advertisers who need it. As pervasive as their own surveillance is, the platforms' data extraction machine runs up against one, forever unsurpassable barrier: They can never truly know our intentions. The data they feed to the online advertising industry is probabilistic and predictive but by no means definitive. With cookies that track your activity, websites might surmise that you're possibly in the market for a new car. But the gap between those signals of interest and the moment at which you write a check is still very wide, which means marketing departments' "customer acquisition" costs remain high. For all the high-tech digitalization and the granular data that help them identify prospects, advertisers essentially use the same scattershot, wide-net approach they've used for two centuries. They put an ad in front of as large an audience as possible in the hope that a small number of those prospects will convert into real, paying customers. In theory, the more precise the data and the more likely a particular audience is to convert, the higher the cost per mille (CPM)—that is, cost per thousand impressions—websites can charge to advertisers. Yet even as they are armed with all these tracking devices, digital media companies struggle to charge high enough CPM rates to make decent revenue because they can't give their advertising clients a high enough level of confidence about their customers' intentions. Both sides are denied access to the platforms' black boxes; both remain in the dark about their audience and customers.

Now imagine you are living in the next-generation internet we've been discussing, one in which you have control over all your digital data. You want to buy a car and are willing to communicate that intention to a range of auto dealers in return for genuine discounts. In this new internet, those dealers have no other way to second-guess your intentions from your online activity because it's protected, under your control, and not available to them unless you give them permission. This puts you in a solid bargaining position. How might you use it? Well, with your new self-sovereign powers you could send your favorite car-themed news site a message that's unique to you and that reveals your purchasing intentions. You wouldn't share your name, your address, or any personal information; you would merely signal that the unidentified person in charge of your unique internet address (i.e., you) is looking to buy a vehicle with certain specified characteristics and that you're open to receiving up to, say, ten offers. In return for this, you lock in a financial credit from the website that you can claim against your future car purchase and then wait for the dealers to contact you. Since they can't know who you are and to whomever else you're talking until a deal is signed, you can comfortably negotiate them down to your ideal price, playing one off against the others, and then claim your credit from the news site on top of that.

In this scenario, you are selling a piece of data—a message that you want to buy a car—to the news site, which in addition to delivering interesting articles on automotive trends acts as a broker of that data to the market of car dealers. The dealers are prepared to pay generously—in effect, paying a higher CPM—for that data, because it reliably conveys a clear intent to buy. With the payments that those dealers make, the auto website builds a pool from which to offer credits to its readers. Everybody wins in the transaction. The car dealers are delivered real, not phantom, prospects. The media company can charge a very high CPM rate for delivering a high-value connection. And you, the buyer, get to harness your valuable data into the best deal you can make.

You could be compensated for your data in transactions like these in any multiple ways: for example, in kind, in direct cash, or in the positive feeling you get from contributing something valuable to a cause you care about. Sharon Terry, the mother of two children with a rare skin disorder known as pseudoxanthoma elasticum (PXE), embodies the latter. In a 2016 TED Talk, she recounted how she discovered that researchers studying the disease never shared data and histories with one another "because the ecosystem was designed to reward competition rather than to alleviate suffering." Frustrated, she and her husband set up a nonprofit that coordinated a volunteer network of people donating blood, tissue samples, and medical histories. They distributed the data to researchers on the condition that they share their work with one another and with the donors. Eventually, their work led to the discovery of a PXE gene, which they patented under an open-source license and made available to all researchers. Such opportunities to make a difference could become commonplace for ordinary people when their existing medical data is under their control in digital form.

It's impossible to predict what use cases will arise in this new data ecosystem, which is what you'd expect in a functioning, open-ended market. It is up to the buyers and sellers of that data to figure out how best to value and exchange it. We believe, however, that if businesses are no longer focused on accumulating troves of data about us for repurposing or reselling, and are instead incentivized within a trusted internet to satisfy their customers, they will come up with endless new products that will create real value for all of us.

Surely this is what's best for a capitalist society. In a truly open system in which no centralized power controls all the information, the multiple actors making economic decisions will collectively activate the hidden hand of the market to fund for-profit solutions that produce the optimum outcome for all. That's when rewards and social outcomes are aligned. Isn't this exactly what was unleashed when, along with our human rights, we were given property rights as part of the American Project?

The New Economics of the New Internet

By re-architecting the internet around a decentralized model, in which control over data and content is in the hands of the consumers and producers who care about it and who are incentivized to produce products that bring true value to others' lives, we'll overcome several other major failings with the current digital economy. In this section, we highlight four: storage economics, data portability and interoperability, cybersecurity, and digital property rights.

Storage Economics

One source of the Big Tech platforms' vast influence stems from a major economies-of-scale advantage: their investment in high-cost data storage, an expensive service that's essential if the members of online communities are to access a common set of information, such as the same news feed in a social media group, maintained in a steady state. The "steady state" requirement means that there must somewhere be a common store of data that community members' computers, including their smartphones, can access at the same time. The bigger the internet became, the bigger the data storage needs became, which is what gave rise to the phenomenon of giant data centers, sometimes called data farms. Now a fixture in underpopulated, dry, steady-temperature areas such as the high-altitude regions of Utah and Nevada, these massive warehouses, filled with fast-whirring servers, are the backbone of the internet. The biggest of them are owned by major internet firms such as Google, Amazon, and Microsoft. Since these were the only entities that had amassed enough money to pay for such capital-intensive operations, they ended up being the internet's data storage providers, leaving billions of users dependent on them for the steady state they delivered. That dependency reinforced the companies' dominant position.

Now, however, the idea of a data farm is redundant. New decentralized storage solutions are moving us to a "shared state" world. With a shared state, responsibility for managing websites' storage and hosting needs is spread across multiple, independent computers, whose owners are compensated for providing the service but unable to manipulate the data. This neutralizes an argument, founded on the Economics 101 concept of "economies of scale," for why the platforms are needed. Previously, this highly costly, centralized storage function could be performed only by big organizations with enough scale to operate and run them. Now it can be a collective, decentralized service in which anyone can participate. That's a game changer and a key reason why, in the next generation of the internet, we just won't need the dominant platforms of the current one. This is a major, and welcome, shift.

Data Portability and Interoperability

One of the most impactful reforms to emerge from the early web era's landmark Telecommunications Act of 1996 was a requirement for the regional "Baby Bell" telephone providers to open up their physical phone-line networks to new, competing telephone and data providers. (My family took advantage of this when RCN, a company my brother David founded in 1993, launched the first bundling service, mixing telephone access with internet and cable TV.) The move had the intended effects of encouraging low-cost data and voice service offerings and of driving innovation in alternative distributions systems such as wireless telephony, right as the internet was taking off. The act essentially established the user, not the company, as the "owner" of the phone number. Practically, it meant that people who left their Baby Bell provider could take their phone number with them and use it with a competing provider. The drafters of the law recognized that, to spur real competition, it wasn't enough to open up the legacy networks to newcomers; they needed to ensure that users weren't disincentivized to migrate to the new providers when doing so would have required that they contact all their friends and business associates to update their information. We can now apply this same thinking to the portability of data and the interoperability of platforms and applications.

There is no 1996 Telecoms Act equivalent for data portability or app interoperability, because the current internet platforms are not designed for it, nor do they want it. Sure, as mentioned in chapter 3, Facebook, Spotify, and other platforms do allow you to download your data, contacts, playlists, and so forth, and third-party providers sometimes offer mechanisms for you to import them into a competing platform. But what's needed is a mechanism for us to connect individually and independently with each other. If the telephone companies simply gave you a phone, complete with a phone number, but gave you no ability to call someone, it would be useless. It's why portability of data must also come with interoperability across platforms and apps.

Right now, the different platforms have no incentive to act as competing telephone carriers were compelled to in 1996, to forge interconnections with one another. They know their value lies in creating "walled gardens" that use the appeal of their network effect to entice people into staying with them forever and delivering reams of valuable, exploitable data in the process to their proprietary algorithms. But, as we've argued, in a world with a decentralized universal social graph in which we control our own data, an open ecosystem of permission-based interconnections, the network effect essentially belongs to everyone. In the future, streaming services and other providers of valuable digital products will instead be incentivized to provide the best experience possible for us, their users, which includes the capacity to bring into their environment the digital records we've accumulated elsewhere. They'll have to compete with other providers to keep us or win us back with higher-quality recordings, lower prices, or new innovations and features. And that's as it should be in a functioning market economy.

Cybersecurity

In a 2022 article, *Cybercrime Magazine* estimated that the total cost of cybercrime to the world economy from "damage and destruction of data, stolen money, lost productivity, theft of intellectual property, theft of personal and financial data, embezzlement, fraud, post-attack disruption to the normal course of business, forensic investigation, restoration and deletion of hacked data and systems, and reputational harm" would run to $8 trillion by 2023. If cybercrime were a country, it would be the world's third-largest economy. Amazingly enough, that staggering sum makes it one of the most successful internet industries ever created. That's hardly a ringing endorsement of the current internet's design principles.

The internet's current design leaves zettabytes of valuable data stored in single, hackable data stores, each controlled by distinct centralized entities. They are enticing targets for cybercriminals. If they can get behind the firewall, accessing that mother lode will pay off handsomely. And many do, repeatedly. Hackers are getting more and more sophisticated, which means that ever more money must be spent by security providers to create ever more sophisticated firewalls that keep the attackers out. For an analogy, think of an enemy army trying to get over a castle wall and constantly adding new rungs to an ever-extending ladder as the castle's protectors struggle to add new bricks on top of the wall but frequently fail to do so in time to keep the attackers out.

It's the economics that make this system so vulnerable: The very fact that these pools of data exist and are constantly growing in value makes them an irresistible target for attackers. After all, the hackers have to find only one weakness in the firewall, one "vector of attack," to gain access to all that the castle has to offer, whether it's a bank's vital customer data, a government's tax information, or your children's school's database.

Here's the thing: A model in which our data is under our own control and stored within a decentralized, shared-state framework is not only socially and economically preferable but also far more secure. The model shifts the economics to the point where it will not be worth the effort for hackers to try to steal it. With DSNP, hackers must expend effort and energy on billions of separate attacks to get to the same mother lode they could previously have tapped by just orchestrating a single breach. You could say we'd "make cybercrime unsexy again."

Digital Property Rights

Until recently, there was no way to identify distinct digital assets. With "copy and paste" or "save as," it was trivial to massively replicate a document, a song, or a video. Unlike a photocopy in the analog world, a digital copy is identical to and indistinguishable from the original. So, without a universal record-keeping system with which to track and prove the origin and provenance of such digital items, there was no way to establish "scarcity" in digital form, making it hard to properly identify something digital as your unique *property*. This led lawyers for large entertainment companies and publishers to dream up the digital rights management (DRM) system, in which we all became licensees of content, not owners. Whereas you could own a physical book in the analog world—but not the copyright attached to its contents—and sell it to whomever you pleased, you could not own a digital thing. The whole matter of digital rights became tied up in a highly litigious and restrictive system that served no one but the biggest platforms, entertainment companies, and a breed of gotcha-seeking lawyers known as "copyright trolls." As the internet behemoths quickly took control of the markets for streaming and other content distribution services, individual artists, musicians, filmmakers, and other creators could no longer independently form direct relationships with their fans.

DRM had direct implications for the business of social media, which was built on user-generated content. We users were not going to hire lawyers to defend against the misuse of the posts we created, even though under long-standing legal principles we would start out as the natural owners of that copyright. So, Facebook, Twitter, and others came up with a simple solution: They got us to sign away all such rights, assuming we'd never be able to monetize them and that we didn't care anyway because, hey, look at all those "likes"! Most important, the DRM framework created a mental model that shaped how tech entrepreneurs and their investors viewed the role of data, which was fast becoming the most valuable commodity in the world. They fed us a line that it was only useful and valuable in its aggregate form, as a pool of behavioral information controlled by the biggest players to be sliced and diced and analyzed in relation to itself. No one even considered that individuals might be able to extract value from their data. That justified what we now see as a breach of our rights:

We have never been allowed to claim our data as our property. Instead, it was controlled by the centralized internet platforms, the feudal lords of the digital economy, not by the people who generated it.

The massive value generated by property rights on the internet is currently divvied up among a few big players. And if we stick with this system, it is going to get even more imbalanced as the age of AI takes off. All the unique content and data trails we've produced, the digital property to which we should have rights, is already being ingested into machine-learning black boxes and then processed and spat out without a way to trace the source of the content.

Thankfully, there are now ways to create and enforce those rights before they are fed into the AI black boxes. We are entering the age of digital property rights. That's driven by two big technological shifts. The first will embrace new online wallet and decentralized identity systems, such as those enabled by DSNP, which will afford each of us control—in other words, a direct actionable property right—over the most important commodity of the digital economy: our data. The second involves forging unique digital assets, by which we mean anything and everything that now, and in the future, can take on a digital form. With specialized marking technology—akin to a certification of an original work of art in the physical world or a serial number on a kitchen appliance—we can establish both the provenance of digital things and their ownership. For the first time, an original digital work can be distinguished from a copy and can be associated with whoever has the right to use and commercialize it, whether that's the creator or the subsequent owner.

History tells us that the advent of digital property rights could generate massive economic expansion. In times when property rights have been extended to a significantly wider class of people and entities, we have seen some of the biggest moments of wealth creation. When, for example, the Dutch East India Company was founded at the start of the seventeenth century, creating the first joint stock company and extending the right to own profit-making businesses to anyone who could afford to buy a share, capitalism took off. Fast-forward to 1982, when Deng Xiaoping finally won the leadership struggle for control of China's Communist government that had been set off by Mao Zedong's death in 1976. One of the first acts of reform that Deng instituted under his "socialism with Chinese characteristics" was to allow people to own and monetize the homes they occupied. That too-little-discussed act is one of the most important contributors to the four-decade boom that would follow, taking the Chinese economy from a backwater basket case to the second biggest in the world.

Both of these historical events are examples of what the Peruvian economist Hernando de Soto has called the "mystery of capital." This is the idea that wealth created in developed countries is a result of sophisticated, trusted legal systems, whereas poorer developing countries struggle to create wealth when their record-keeping is untrusted and where contract law is enforced capriciously. De Soto has become a tireless advocate for developing countries to improve the attachment of legally defined property rights to the assets people otherwise rightfully own. It's the way to unlock what he calls "dead capital" and create wealth for many.

Let's extrapolate from de Soto's thesis to imagine how an online economy might grow and evolve if everyone is granted digital property rights. If we hand people the power to make something of the digital assets they accumulate through their online activity, from their creative output, and with the information in their social graphs, who knows what value they'll create for themselves and for the world?

~

So, we've now mapped the positive-reinforcement mechanisms to transition from an unhealthy society in the digital age to a healthy one. As the preceding chapters have laid out, it's about rearchitecting the internet so that it embeds the values that such a society aspires to maintain. It's about enabling the appropriate *rights, responsibilities,* and *rewards* for people to be their best selves and pursue their self-interest in ways that align with the public interest. Sometimes, however, people will fall out of line; that's why we need governance. Like it or not, we need *rules.* That's next, in chapter 5.

CHAPTER FIVE

Rules: Governance in the Internet Age

Society is produced by our wants, and government by our wickedness; the former promotes our happiness *positively* by uniting our affections, the latter *negatively* by restraining our vices. The one encourages intercourse, the other creates distinctions. The first is a patron; the last a punisher."

So wrote Thomas Paine in *Common Sense.* Paine addressed this "necessary evil" aspect of government after first asking his readers to imagine an early, isolated utopia with "a small number of persons settled in some sequestered part of the earth." He contended that "in this state of natural liberty, society will be their first thought," not government. Paine suggested that in such a proto-society each inhabitant would soon recognize that communal labor and support would prove mutually beneficial. In this initial phase, he argued, the "blessings" of this new society would "supersede, and render the obligations of law and government unnecessary while they remained perfectly just to each other." Conceding, however, that "nothing but heaven is impregnable to vice," he warned that eventually this new society would need to establish some form of government to keep its people's worst instincts in check.

It is with a similar sentiment that we've chosen to place *rules* as the final of our Four Rs. We needed to first address the *rights* of the individual, the *responsibilities* of citizens, and the *rewards* and incentives that should guide economic actors. Taken together, these establish the virtues that underpin any great society. But, yes, people are going to do bad things, including on the internet. The last thirty years offer rather compelling proof of that. We now know it was a grave mistake to assume that people would naturally behave well toward one another on the internet. So, unavoidably, we need to think about governance. By that, we mean the system for establishing and enforcing rules of behavior, a concept that can include governments as they are traditionally known but that can also include a variety of self-regulated models.

Paine avoided being overly prescriptive when he turned to the topic of governance in *Common Sense.* It was not up to him, he said, to define the model; it was up to the American people collectively. He emphasized the principle of self-governance, which in practical terms meant states and local communities should be empowered to a great extent to govern themselves. That decentralizing ideal, known as federalism, became integral to the design of the American Project. To be sure, in its first decade the United States leaned a bit *too* far in that direction. The first constitution—the Articles of Confederation—left too much power in states' hands and gave Congress no real authority, which allowed division and dysfunction to fester. Nonetheless, decentralization and self-governance remained as core tenets of the Constitution of 1789. That's why, to this day, even in an era when the highest court in the land is politically divided, lawyers who bring cases against the U.S. government to defend states' rights against federal overreach continue to find sympathy among the justices of that Supreme Court.

What if we applied the same decentralizing principles to the next generation of the internet? Our argument that we should make a technical change to the internet's architecture by adding a new, human-centric protocol is akin to proposing a constitutional amendment to right an imbalance in government, correct a wrong, or stay current with the times. Like the U.S. Constitution—on which so many other laws, judicial systems, executive agencies, functions, and institutions have been built—the base-layer protocol should be a thin piece of code. It should have limited scope, establishing the foundational rights of the people who use the internet, and it should allow developers to build a wide array of mid-level protocols, services, and applications with their own specific rule sets on top of it.

Right now, the "government" of the internet is in the hands of the companies that run its major platforms. They act as gatekeepers, deciding who can participate, what can be said, and who gets to listen. But must it be that way? Imagine what governance for the internet might look like if it was in the hands of the people. Under the DSNP model we've laid out in prior chapters, self-governance arises as people achieve control over their data and content. Thus empowered, they can band together to form online communities of their collective making and then agree on what information can be shared to the group and who gets access to it. To do this, the group can be narrowly empowered with the same permissioning capabilities that DSNP provides to individuals. Helped by a variety of new decentralized governance tools, the members can meet, vote on proposals, agree on moderation of content, and so forth, all in the interest of ensuring that the group's rules are established and enforced with a sufficient degree of democratic oversight.

The principle is that it's up to the members to decide among themselves what information is admissible to the group; it's not dependent on the curation and censorship of a giant tech company. Likewise, apps built on DSNP can establish provable policies and practices around their content and other offerings—regarding censorship or moderation, for example—and people can choose to interact with those apps or not. If the apps they choose don't live up to their promises, those users are free to move, along with their valuable social graph data, to some other provider.

But before we explore ways in which internet self-governance can be applied, we must look more deeply at how its autocratic structure has undermined the real-world democracy that the Constitution was designed to preserve.

Accountability Lost

You've likely encountered at some point in your life these inspiring words from John F. Kennedy: "We choose to go to the moon in this decade and do the other things, not because they are easy, but because they are hard." Delivered in a speech at Rice University in September 1962, they're often cited as an example of a bygone America, a can-do America that embraced challenges, an America that saw technology as a driver of progress and viewed the future with hope and wonder, an America in which our government seemed to work reasonably well. (In this book's conclusion, we'll offer our vision for how this America might be revived.)

Another way to view the "We choose to go to the moon" speech is as an investor pitch. Kennedy needed to convince his fellow Americans to fund this audacious project. Americans who had lived through the Great Depression and the Second World War would have seen Kennedy's moonshot as extravagant. Indeed, over thirteen years, the Apollo project would cost taxpayers a total of $25.8 billion, the equivalent of $300 billion in 2023 dollars. So, at Rice, Kennedy had to be up front with the money part. Noting that the 1962 space budget would be greater than the previous eight years' budgets combined, he warned it would grow to 50 cents a week for each American. But then, like an inspired start-up founder closing out a presentation to a venture capitalist, he made the case that it was all worth it. In a sentence that clocked in at 159 words—a rare case of an effective sentence of such length—he described the intricacy and thus the enormity of the "hard thing" he was proposing, thereby rallying his audience to commit to the cause:

> But if I were to say, my fellow citizens, that we shall send to the moon, 240,000 miles away from the control station in Houston, a giant rocket more than 300 feet tall, the length of this football field, made of new metal alloys, some of which have not yet been invented, capable of standing heat and stresses several times more than have ever been experienced, fitted together with a precision better than the finest watch, carrying all the equipment needed for propulsion, guidance, control, communications, food and

> survival, on an untried mission, to an unknown celestial body, and then return it safely to earth, re-entering the atmosphere at speeds of over 25,000 miles per hour, causing heat about half that of the temperature of the sun—almost as hot as it is here today—and do all this, and do it right, and do it first before this decade is out—then we must be bold.

Now fast-forward six decades. Try to imagine another world leader, Vladimir Putin, making a similar speech to his "fellow citizens" in Russia to support his invasion of Ukraine. You can't. Putin has never had to "pitch" Russians on anything. This is why democracies, not dictatorships or absolute monarchies, are breeding grounds for political oratory, a skill that, done right, will go beyond merely selling an idea to an audience and extend to inspiring an entire nation into committing to a cause. (Think Lincoln's Gettysburg Address, Churchill's "fight them on the beaches" speech, Franklin D. Roosevelt's "fear itself" classic.) Whereas a dictator or an absolute monarch has carte blanche, a president must work hard to secure the support of the citizenry. In a democracy, there is no "power" (*kratia*) without "people" (*demos*). It's the essence of popular sovereignty.

Okay, now let's consider the words of a more recent American "leader": In 2004, after starting what he originally called Facemash—a site assessing Harvard students' looks, primarily intended for men to judge women—Mark Zuckerberg was reportedly asked by a roommate why people willingly submitted over four thousand emails, pictures, addresses, and student numbers to him. He replied, "I don't know. They trust me. Dumb fucks." Perhaps this should have been a warning sign, a hint at the exploitative nature of what would later be created. Worldwide, more than 3.5 billion people—more than three-fourths of all internet users—are on one or another of Meta's platforms. Maybe Zuckerberg's comments were prophetic.

If it were ever discovered that the CEO of a construction company had such contempt for the people entrusting it to construct a building or an important piece of infrastructure, the company would never again win a contract, attract investors, or retain a client. By contrast, you find the feudal lords of the leading digital platforms doubling down on their model of data extraction and manipulation. That's because there's no effective governance overseeing those business executives, one that demands of them the kind of accountability Kennedy was held to when he pitched his big ideas to the American people. The executives and their companies are, in effect, above the law. Since they, and only they, control the data and the algorithms that determine our behavior, they get to set the *rules*.

Two decades after his Harvard days, with Facebook, Inc.—or, under its new name, Meta Platforms, Inc.—Zuckerberg and other internet CEOs continue to operate as if they own the social connections we form and can do with them as they please. In the summer of 2023, Elon Musk, the owner of X (at that point still called Twitter), literally challenged Zuckerberg to a mixed martial arts duel, accusing the latter of stealing X's intellectual property—by which Musk meant the data that X had accumulated on us, *our* data.

Zuckerberg, Musk, and their ilk are feudal warlords, fighting over the territory they've seized from innocent people. We are not even afterthoughts in their battles. We must start viewing them this way and be done with their models. Fortunately, if we all migrate to a new system in which we each have autonomy over our data, we can collectively reset the internet to one that "We the People" govern. And we can do so without resorting to the violence that America's founding fathers were compelled into. We won't even need the intervention of our ineffectual government representatives, who've shown themselves to be too out of touch and too corrupted by Silicon Valley money to act on our behalf.

While none of us has an excuse for ignorance anymore, don't beat yourself up over failing to see the exploitation perpetrated by these platforms in years prior. Pretty much everybody missed the forest for the trees. Preexisting legal and popular definitions of what constitutes a private enterprise versus a public entity obfuscated the problem at hand. We blithely accepted the internet titans' contention that, as private companies, they weren't bound by the constitutional obligations imposed on governments to protect free speech, privacy, and other rights.

From time to time, certain critics of the platforms—including elected officials who've been upset over having their accounts censored or being "deplatformed" altogether—invoked a longstanding debate over Section 230 of the 1996 Communications Decency Act. If the platforms are going to claim a private right to discretion over censorship and curation, these critics argue, the companies can't simultaneously enjoy Section 230's exemption from liability over everything their users say on their platforms even when they choose to moderate some of their more offensive content. (Section 230 treats internet platforms like newsstands: Whereas the publications that a newsstand displays on its shelves can be sued for libel over the content they publish, the newsstands themselves cannot be held accountable for libelous content and can choose to display the publications however they please.)

We see the fight over Section 230 as a distraction. It does nothing to address the more urgent human rights concerns that stem from the platforms' abuse of our data and our personhood. Without a fundamental overhaul of this model and a reassertion of people's civil rights within it, any repeal of that provision could actually have the chilling effect of curbing free speech on the internet.

A Captured Washington

Some argue that the way to bring the Big Tech companies to account is to regulate them as public utilities. Whether we like it or not, social media platforms have become society's information system. They've seized a centralized gatekeeping role in the distribution of information, an essential services function that can be compared to that of public utilities such as electricity, water, and toll road providers, which are tightly regulated. Power companies are typically required, for example, to deliver electricity at the same price to city and rural residents, regardless of the extra cost of servicing the latter. Wireless providers, whose licensed access to the public radio spectrum is at the privilege of the government, are required to fulfill a range of functions in the public interest, such as providing emergency communications. Why not demand that the internet platforms hold a similar degree of public responsibility? Others invoke the same argument to push for the use of antitrust laws, those that President Teddy Roosevelt wielded in the early 1900s to curb the power of Rockefeller, Carnegie, Vanderbilt, and other robber barons. Can we just designate Meta,

Amazon, and Google as robber-baron monopolies and break them up? The answer: Maybe, but it would not solve the problem. The power these entities have—quite literally to shape how we think—is more harmful than anything any service monopoly has accumulated in the past. It cannot be resolved by simply requiring them to do this or that thing differently. And if we fragment them into smaller platforms but let each of the constituent parts retain control over the social graph, they will simply use that control to rebuild their dominant position. No, we must strip them of unfettered access to our data and the opportunity to exploit it against us. The other problem with this approach is that it relies on action from a dysfunctional Congress that seems incapable of legislating—partly because it has been captured by Big Tech.

Roosevelt recognized that extortionist economic power in the hands of monopolies results in bigger problems than just higher prices for consumers. In fostering society's dependence on their services while accumulating a war chest of funds with which to lobby politicians, monopolies also wield outsize influence over the nation's political process. We saw this with Wall Street during the 2008 financial crisis, as "too big to fail" banks held the government hostage, extracting bailouts for their shareholders in the name of saving the financial system. We have the same, arguably even bigger, vulnerability with Big Tech.

From time to time, Congress puts on a show of scrutinizing Big Tech. Mark Zuckerberg became a whipping boy for grandstanding lawmakers in 2019, when Facebook's plans for its Libra cryptocurrency prompted a backlash from governments worldwide, and then again in 2021, when the whistleblower Frances Haugen delivered damning testimony about the company's abusive practices. And, as mentioned in prior chapters, the Department of Justice and state attorneys general have lately, finally, become more aggressive with lawsuits. There's also decent lawmaker support for the Kids Online Safety Act, which Project Liberty has supported as an imperfect stopgap measure—until the entire system can be redesigned—to hold the platforms accountable for putting children in harm's way.

These are worthy efforts. Observers are calling the most recent Google case the most important antitrust action since the Department of Justice took on Microsoft over the Internet Explorer browser. Coupled with the Epic Games victory, it suggests a tide of public and judicial opinion turning against the platforms. And while fines in the past have been paltry, the numbers now are expected to become more sizable and painful for Big Tech to absorb. Yet, given that they have gargantuan war chests, what will this judicial approach actually yield? Unlike with the landmark cases against Big Tobacco, when society benefited from the pressure to reduce the harmful effects of smoking, we don't want people to stop using the internet. We want to fix the internet. And if all we're doing is making the platforms pay up for the right to continue their closed, oppressive system, nothing is achieved.

People who reflect on Washington's track record can also be forgiven for doubting it will act in the public's interest going forward: The U.S. government takes actions to protect the platforms' monopolies. When the European Union has pushed through legislation and lawsuits aimed at curbing their power, different administrations have often come to the companies' aid in the fight with Brussels, all in the name of defending "U.S. interests." (You rarely hear government officials entertain the idea that the platforms have forfeited the right to be associated with our interests by manipulating American citizens' minds.)

The truth is, in Washington, money talks. And with Alphabet, Amazon, Microsoft, Apple, and Meta commanding a combined market capitalization of $10 trillion when this book was going to print, the platforms have bucketloads of talking power. Also, whereas the manufacturing industry exercises influence both by dropping cash into politicians' reelection campaigns and by promising jobs to union members and other people in lawmakers' constituencies, internet platforms have found additional, more subtle ways to cozy up to them. The revolving door between Washington and Silicon Valley is starting to look like that between Washington and Wall Street, where regulators, lawmakers, and their aides frequently go on to jobs in finance at the end of their stint in government, creating an uncomfortably tight relationship between these institutions and the agencies assigned to regulating them. In just a few months in 2023, the newest player in the pack of Silicon Valley's data-hogging companies, Sam Altman of Microsoft-backed OpenAI, "stormed Washington" to meet with more than a hundred lawmakers and to visit senior White House officials, according to *The New York Times*. Take a look at Big Tech's workforce, and you'll invariably find high-level positions such as general counsel or chief policy officer occupied by former government employees.

Even more troubling, these companies have directly helped politicians' reelection efforts. Senior employees of internet platforms have been involved in the campaigns of powerful politicians from both U.S. parties. Barack Obama's 2008 presidential campaign was one of the first to employ social media to build a voter base, a strategy that was instrumental in his victory and, for a time, afforded him a reputation as a cool, tech-savvy politician who could reach younger voters through the internet. Presidential candidates and other leading political figures of all persuasions have since followed that lead, helping to normalize manipulation of voters. This algorithm-centric approach, which reduces voters to data points and instincts of actionable behavior, dehumanizes us and undermines the principles of popular sovereignty on which our democracy depends. By 2022, Obama appeared to have had second thoughts about the value of the internet platforms that had supported him, telling a Stanford audience that disinformation online posed a major threat to democracy.

Politicians feel like they have little choice but to participate in social media. They feel pressure to manage their "brands" and get votes, treating their feeds on X and other platforms as de facto billboards, a means of attracting attention and of standing out. Some are more effective at getting attention than others, a skill set often at odds with the unglamorous, time-consuming work required to negotiate legislation, get bills passed, and meet with voters in person.

Former president Donald Trump is, arguably, the grand master of social media. He long ago realized he could garner attention among the most social-media-engaged conservatives by staking out provocative positions on Twitter and elsewhere. But others have become adept at it too. When you have a moment, look at how many followers are on the social media accounts of politicians who occupy the most extreme positions on either side of the aisle and compare those to the accounts of more moderate, middle-ground-seeking elected officials and candidates. You'll find that the former attracts a far bigger following than the latter, which has a highly unhealthy impact on civic discourse and on the practice of government, leading to unbalanced political outcomes or no outcomes at all.

And the United States is far from the only place in which this is happening. In fact, it was an overseas election—the UK's Brexit referendum—that most clearly revealed this technology's harmful impact on democracy. Thanks to *The Guardian*'s Carole Cadwalladr, who reported on Facebook's role in the Leave campaign's success, we're now aware of the concept of microtargeting. After the United Kingdom voted to leave the European Union, Cadwalladr visited Ebbw Vale in South Wales, a former mining town that had attracted hundreds of millions of pounds in EU funding for a new educational institution, several highways, and a sports facility. She wanted to know why, despite that economic injection, a majority of its residents had voted to leave. As she recounted in an influential TED Talk from 2019, many of the townsfolk told her they were "fed up with all the refugees." That line left her baffled; it was something she thought she'd be more likely to hear in a Conservative voting district than in a staunchly Labour former mining town with one of the United Kingdom's lowest rates of immigration.

After Cadwalladr published an article exploring why the people of Ebbw Vale had inexplicably voted against their interests, a woman from the town contacted her to tell her about all the "scary stuff about immigration and especially about Turkey" she'd seen on Facebook before the election. Cadwalladr tried searching the platform for those posts but couldn't find anything—"because there's no archive of what ads people see or of what had been pushed into their news feeds—no trace of anything. It had gone completely dark." When it dawned on her that individuals in marginalized towns like Ebbw Vale had been explicitly targeted for their potential to respond to such ads, whereas the mostly "Stay"-voting residents of London, like her, were not,

Cadwalladr uncovered a story that showed how "this entire referendum took place in darkness, because it took place on Facebook."

Cadwalladr's and others' reporting shone a light into that darkness when they revealed that the political consultancy Cambridge Analytica had worked with the Leave campaign to systematically spook individuals in certain regions of Britain into voting for Brexit. Further light came from testimonies in Parliament by the former Cambridge Analytica employees Christopher Wylie and Brittany Kaiser, the latter of whom now heads up the aptly named Own Your Data Foundation. It was also later revealed that Cambridge Analytica played a similar role for supporters of Donald Trump during the 2016 presidential election. Records show the company boasting to its GOP clients that, via its Facebook relationship, it had obtained more than 7,000 personal data points on each of 240,000 American voters.

Cambridge Analytica later collapsed, and its name is now infamous. But Facebook lives on. And power-hungry people continue to use it to manipulate others—sometimes to commit the most atrocious acts. In a 2022 report, Amnesty International detailed how, during Myanmar's ethnic-cleansing push against its predominantly Muslim Rohingya minority, the country's brutal military regime used Facebook to help it coordinate "a targeted campaign of widespread and systematic murder, rape and burning of homes." The human rights advocacy organization is now calling on Meta to pay reparations. "While the Myanmar military was committing crimes against humanity against the Rohingya, Meta was profiting from the echo chamber of hatred created by its hate-spiralling algorithms," said Agnès Callamard, Amnesty International's secretary general. "Meta must be held to account."

So, let's summarize what we have learned. The internet platforms' algorithms have helped dictators incite pogroms against minorities, were used to stoke xenophobic fear among British Brexit voters, and continue to encourage politicians in the United States to sow anger and division instead of valuing compromise and effective governance.

Should it be any surprise, then, that digital feudalism has also fostered digital corruption? We must end this madness. We must restore a more democratic system of governance for the internet, one that lives up to Rousseau's model of popular sovereignty to reflect the will of the people. *We* need to set the rules, not Big Tech. Let's apply the values of the American Project, whose system of government was structured to fit the aspirational needs of its citizens and, by enabling them to pursue their goals in concert with one another, help them construct a better world.

Going Our Own Way

One way to enforce change would be to try to convince our elected representatives to enact laws curbing the Big Tech platforms' power. To be effective, these laws would have to do more than simply impose a few new rules requiring the platforms to play nice while keeping intact their core system of data monopolization. Legislation would need to be sweeping. It would have to rip the guts out of their core model. Well-meaning reformers have tried a piecemeal approach to mending our broken internet, but that approach hasn't gone far enough. We've seen legislative initiatives intended to preserve people's privacy (the 2016 General Data Protection Regulation and the "right to be forgotten" laws in Europe, and similar laws in California), to protect children (the U.S. Children's Internet Protection Act of 2000), and to force platforms to compensate news outlets for profiting from their content (in Canada and Australia). But because of the network effects they enjoy, the platforms, as well as the publishers and advertisers that use them for reaching audience, have found ways around these inconveniences. Since the default for everyone—publishers, advertisers, and users—is still one of platform dependency, solutions such as the GDPR-compliance disclosures and opt-out buttons have just become an annoying hindrance to using that system. The vast majority of users just click through them to reinstate the tracking mechanisms the regulations are supposed to discourage. At best, these laws have made people only marginally safer online; mostly all they seem to do is make navigating the web a more unpleasant experience.

We would need a truly foundational bill, one even more sweeping than the landmark Telecommunications Act of 1996, which broke down the Baby Bells' regional telephone monopolies and set up the competitive market for data and telephony, to fast-track the rollout of broadband and the World Wide Web. Much as telephone companies were then compelled to allow competitors to provide services on their network and to allow customers telephone number portability from one carrier to another, this bill would need to mandate interoperability between internet platforms. It would also have to require that they transfer full control and rights over past and future data to the individuals to whom it pertained. They could no longer persist in their current, walled-garden state. Just as you can take your phone number to any provider, you would be able to seamlessly transport your data (your contacts, your playlists, your message feeds) across platforms and apps to connect with whomever you please. The underlying code in their algorithms would need to be open-sourced, too, so that observers could prove that they're not abusing whatever data they continue to glean from new activity. All of this should be enshrined in a law that elevates and protects your right to control who gets to see and use your data.

Sadly, achieving something so comprehensive seems like a pipe dream. A bitterly divided Congress led by politicians with no tech savvy—and fraught with all the captured interests and corruption described above—has little to no appetite for getting into the weeds of a complex legal overhaul that would dredge up all sorts of conflicts. More important, reformists would face the most powerful lobbying army ever amassed. As we discussed earlier in this chapter, Big Tech is well ahead of us on this one. Color us cynical, but the platforms have so insidiously worked their way into politics that it makes little sense to choose a reform pathway that starts in Washington.

Don't get me wrong. I'm all for a government that keeps us safe, supports vital infrastructure, protects the truly needy, and enables effective healthcare and education systems to flourish. And, as we'll discuss in the next section of this chapter, there are certain things the government is going to have to do if we are to optimize this next-generation internet. However, if we're stuck with the current, chaos-creating internet, it's very difficult for any government to be effective in any of those areas. In this dysfunctional environment, valuable taxpayer dollars are squandered. If we fix the internet and its underlying technologies, however, government will not only be more effective but also cost less to run.

With that in mind, let's turn to the other option society could take to upend this abusive model, the one that's less dependent on our broken political system. This path involves a rather simple maneuver that would beat Silicon Valley at its own game. The idea is to use a new network protocol, such as DSNP, along with new self-controlled identity and data management technologies, to collectively engineer a prosocial version of what the Silicon Valley pontificator Balaji Srinivasan described in 2013 as the "ultimate exit." Srinivasan was talking about libertarian technologists using their inventions to simply opt out of the government overreach they despise—Peter Thiel's now-abandoned dreams of "seasteading" in international waters, Elon Musk's plans to independently colonize Mars, and Srinivasan's own proposal to found politically untouchable, geographically unbounded "network states" online. In our case, we need not worry about the cost of rocketing settlers to another planet or of building an inhabitable city on a platform in deep-ocean water. And we're not escaping taxes or declaring some government-free zone. In fact, the powers that be that we are calling out are not governments per se; they are the tech titans themselves.

We need people to decide to migrate to new systems of individual data control and network identities. That way, together, we can kick-start a new, human-centric internet. As we've been saying, technology provides us a way out of digital feudalism. But the technology, though complex, is not the hard part; it is being built as we speak. The challenge lies in getting all of us to take the next step and adopt it.

To spur such changes en masse, we must confront the chicken-and-egg challenge of network effects: People aren't incentivized to migrate to a new network when all their friends, followers, and contacts remain on the old one. The value of any social media platform under the current model largely flows from the size of its user base. Since any new social media tool will, by definition, have a small number of initial users, its inferior network compared with the existing system poses a scaling challenge. By contrast, companies that already enjoy a network effect in one area of service have an easier time creating instant buzz and rapid adoption of a new service. That's why Threads, Meta's competitor to X, gained 100 million users in its first week of operation, when it was immediately made available to the 1.2 billion users of Meta-owned Instagram.

Having acknowledged this problem, now let's find a way around it. One way, once again, is to lean into the power of DSNP. As mentioned elsewhere, a cool, interoperability feature of the protocol is that it allows people to reach out to one another and connect wherever they please, on platforms, off platforms, and, most important, across platforms. That will neutralize those companies' network effect advantage. We no longer have to worry about staying in the places where our friends are if we can interact with them anyway.

That said, we really want this movement to happen quickly, so we must confront the apathy and inertia that often hinder people from changing behaviors and adopting good habits. To overcome them, we should try to learn from the many instances throughout internet history when a new stand-alone service enjoyed rapid mass adoption out of the gate. If the product is appealing enough, explosive growth is always possible in the internet age, since delivery by software is virtually free, instantaneous, and unbounded by geography. It's the "killer app" idea that founders of every start-up promise as they try to convince early investors that they'll generate exponential growth in users and revenues. Heck, the big platforms of today had to start somewhere. And while the biggest continue to exploit their network advantage, much as Meta did with Threads, they do periodically face new, fast-growing challengers such as Snapchat's brief assault on Instagram and, after that, the even more formidable TikTok. Our mission is to generate similar buzz and enthusiasm, not for a new profit-making app but for a free, open protocol that anyone can build on and where no one gets to own the collective social graph and its network effect. It's not a killer app; it's a new killer ecosystem.

One way to build our own network effect is to seek out the help of friendly businesses and institutions that already have their own sizable networks. If an app with millions of existing users chooses to shift its entire user base over to this new model, change can happen rapidly. Or what if companies that want to both do the right thing and seek new opportunities in the open-data economy started insisting that all their providers comply with these standards? Imagine some of the biggest consumer brands—those with hundreds of millions of customers worldwide—driving a mass migration. Such corporate actions can be a force multiplier that accelerates real, meaningful change.

Of course, if we are going to encourage people to use the new system, we can't employ the same secretive, mind-manipulating enticements with which the social media platforms keep people engaged. But that's not to say we can't draw on some of the science behind their power—for positive, prosocial ends. The Stanford psychiatry professor Anna Lembke, whom we met in chapter 2, talks about how the same desire for dopamine hits that leads to substance abuse can also get people to exercise or to appreciate fine art or to laugh and embrace the love of family and friends. In short, dopamine is a motivator. Tapped in the right way, it can drive us to do great things.

An inspiring, rousing speech—like, say, Kennedy's Rice University "to the moon" pitch—is one way to unlock a dopamine hit. People are motivated by a great story; it can get into their chemistry, their psyche, and make them do bold and good things. This is why the real task of getting internet users to give up on the network-effect advantages of an otherwise broken system and instead migrate to a better one falls not to the technologists who are building that better model but to creatives, to artists, to storytellers, to those who can spark the imagination of all who hear the message. Throughout human history, storytelling has been a powerful motivator. People presented with a compelling story often take on challenges and, by extension, thrive. It's what Thomas Paine did with *Common Sense.* It's also seen in Hollywood's pervasive, global influence and in America's daring architecture, its music, its theater, and its modern art. All, in different ways, have helped inspire people over the years to embrace new ideas, to step into the unfamiliar in an effort to find a better way.

This book is an attempt to follow that tradition. But something as monumental as a shift to a new internet design needs more than one book. We need creative people to join us in this great storytelling effort. Paine was a highly influential player in helping foment the American Revolution, but the shift of sentiment in 1776 was really the work of multiple writers, of Thomas Jefferson, of Benjamin Franklin, of John and Abigail Adams, of Alexander Hamilton, of James Madison, of John Jay. It also came from the stories told and retold by locals—stories that worked their way into lore—about the heroism of George Washington, of Paul Revere, of Nathan Hale. Fortunately, the inspiring, hopeful story of change that we want to unleash among you and everyone else requires no bloodshed. All that's asked of you in this modern act of rebellion, this push to throw out our digital feudal lords, is that you be a nonconformist. You'll be outside the crowd, distinct from those who persist within Meta, Amazon, and Google's fiefdoms, until they too catch on that the new world of data autonomy is where they need to be. It's how all revolutions start: spearheaded by those who are most impassioned by the cause, a group whose actions inspire others to join so that, eventually, they are unstoppable.

We want you, like the eighteenth-century American revolutionaries whom Paine helped inspire, to be driven both by an angry determination to overturn the unjust current system and an enthusiastic desire to build a better society once that first step is achieved. Part of that bright promise is found in an opportunity to revitalize democratic self-governance in the internet age and, by extension, to empower individuals and communities. In this new era of the internet, we the newly re-empowered digital citizens of the web, will get to define the rules and protocols for how we govern ourselves.

A New, Democratic Gatekeeping Model

The building block for online community self-governance is the individual, whose claim on the first of our Four Rs—*rights*—affords them autonomy to determine who has access to their data and content. Importantly, that autonomy includes the power to revoke access that was previously granted, a feature baked into DSNP that makes it very easy for auditors to check that a site has scrubbed your data. In effect, it makes Europe's "right to be forgotten" laws actionable. These kinds of powers mean that we individuals, not the platforms, will be the arbiters of the new internet's information flow.

When we join an online community as self-empowered individuals, we will grant limited informational access to a group of people who have chosen to abide by that community's mutually agreed-on set of terms and conditions. Under a DSNP model, we and our fellow members can then form a collective entity with an internet identity of its own to represent our common interests within the boundaries of the group's narrowly defined purpose. After setting the entity up with DSNP-enabled tools, we can now use the same permission-granting technology to imbue the entity with its own gatekeeping powers, including authority over who gets access to the group's shared information. What precise powers the common entity receives will vary according to the purpose and mission of the group. Is it a "friends and families" notice board whose members are dedicated to sharing photos of one another? Is it a collectively managed marketplace for trading assets? Are we, the members, identified by our real names or by pseudonyms and/or avatars? In keeping with our individual rights, each member can choose to leave the group by revoking the access they'd previously granted to it. Equally, the group has its own revocation power—ideally wielded within a democratic and trusted framework in which members have some degree of voting power—over any individual found to be in breach of the terms and conditions. The group might also be granted the power to make case-by-case determinations around one-off pieces of content. A designated moderator of a DSNP-based social media feed could be empowered to block certain information from a provider, but not from others, if the provider is in breach of the rules.

Some people might call this information-gatekeeping "censorship"—a lightning-rod topic in any discussion about the internet. But before jumping for such labels, note the different power dynamics between this internet model and that under which we currently labor. In our vision of the future, we the people will have much more agency, through a system of shared governance, over what information we choose to receive. In the current system, we have no control over, or even insights into, the secretive process by which Big Tech controls information. In the human-centric internet, people will choose to join groups of others who share their views. There, they will collectively decide what content is acceptable or unacceptable. As is the case now, that will still mean that online communities arise whose conversations you might find distasteful, prudish, radical, or conservative. Such are the open-boundary requirements of a right to free speech, a vital companion to the right to own our data. But in wresting back control over our neural pathways, over our free will, we'll now be able to more clearly choose to avoid such content if we so desire. We simply won't join the groups we don't want to belong to.

Right now, by contrast, we are at the mercy of a few social media companies' black box algorithms, which select from a vast universe of content that they've scraped from our creative work to deliver us material that's expected to drive engagement for their ends. Facebook's underpaid, sleep-deprived moderators are no match for clickbait-seeking social media posters who, working in tandem with these engagement-farming systems, get around them to slide an array of racist, misogynist, judgmental, bullying, reductionist, untruthful content into our increasingly toxic online environments. At the same time, those same moderators have over the years been the subject of countless controversies for some ridiculous deplatforming decisions. (*The New Yorker* was once barred from Facebook after it posted a simply drawn cartoon by Mick Stevens of a topless Eve, with black ink dots for nipples, talking to Adam in the Garden of Eden.)

The bottom line is, we must never deny anyone their right to free speech. But while protecting this important standard of "freedom to," we now have an opportunity to reinforce a "freedom from." If we can empower people and communities to choose not to subject themselves, or their children, to online toxicity or cyberbullying, we believe we can foster a more empathetic, mentally healthy, constructive society. Extremism will always exist, but if the platforms and their users aren't consistently rewarded with more engagement for publishing polarizing material, that material will be denied the oxygen it needs to survive.

It will be important in most cases, too, that these new communities' sites are based on open-source code. The lack of transparency in the proprietary, black box algorithms employed by the big internet platforms makes it impossible to judge the integrity of the curation system, whether it is supporting the interests of the group or protecting some special private interest. Future online communities should demand that the code they run is visible to all. This will be doubly important in the AI era. When AI agents are doing everything from managing our investment portfolios to running our healthcare devices, society must be able to investigate the code and the source of the data they're trained on. Obviously, we'll need trusted experts to perform these tasks and report their findings to the nonprogrammers among us.

That brings us to another issue. The matter of content moderation and other policies is, of course, relevant not only to these self-governing social media communities but also to private companies and other for-profit entities that will build apps and provide services in this next-generation internet. How will they be governed? How will their policies and practices be judged and assessed?

Here, too, we can conceive of models that won't heavily rely on the regulatory powers of traditional government. We can start by borrowing from some of the nongovernmental and self-regulating approaches that have developed over time in the United States and other democracies. Think of the role played by private, for-profit auditors of public companies, by self-regulatory organizations (SROs) that coordinate the membership of professional bodies, and by quality assurance mechanisms such as the ISO 9000 set of international standards used in the certification of businesses and their products. Other examples include the Good Housekeeping Seal, awarded by *Good Housekeeping* magazine to products that passed rigorous standards so that consumers could confidently buy, say, a vacuum cleaner that would not spark a house fire. Underwriters Laboratories did the same for all kinds of mechanical and electronic devices, and we're all familiar with groups that certify the safety of toys, baby products, and other items to keep our children free from harm. While there may be a government-mandated requirement in various cases for a firm to be audited, to join an SRO, or to earn a certain quality standard, what's most at work here is the power of the market. The idea is that when customers make a choice between a provider with one of these seals of approval and one without, they should gravitate to the former, all things being equal.

In a next-generation internet, these kinds of approaches for holding service providers to account will be all the more powerful. When our data is in our control, when software is based on open-source code, and when distributed ledgers generate trusted records of data provenance, the investigation process required for certification becomes much more efficient. Do you need assurance that a ride-sharing app isn't accumulating and packaging information about your trips to and from home? Well, a trained auditor should be able to easily verify that. They would check the company's logs showing how information pertaining to each uniquely identified user traveled through its system. If a ride-sharing company chooses not to share such information, even though it could be easily compiled from DSNP-enabled accounts, it won't earn the stamp of approval you're looking for—and you won't use its services.

We have one important caveat to all this: To achieve full data autonomy for the next generation of the internet, we will need some help from government. The biggest demand we'll have is for a law, and for the enforcement of that law, requiring existing internet platforms and apps to hand over to us the archives of our individual and social graph data from the second-generation era in a form that's actionable. It's all very well for us to seize control over all the data we'll generate henceforth after setting up a wallet and starting to use DSNP-enabled sites—and there'll be nothing stopping you from doing that immediately. Also, recall, DSNP's interoperability features will allow us to keep engaging with one another even if we leave our old data under those platforms' control. But to fully regain control over our personhood, we'll want to access all our past data as well, and, absent massive public pressure, that may well require the government to step in and force them to give us what's ours. The problem can't be resolved by technology alone.

A related matter is that to fully enable all features of personal data management under DSNP, the capacity to revoke a service provider's previously granted access to our data will require some kind of enforcement by government or a self-regulatory body. However, as discussed, the proofs required to ensure that data has been properly scrubbed should be fairly trivial to create since the data will have full provenance associated with it and with our DSNP addresses. There should be no ambiguity in that case—unlike with the archival data accumulated under the old web model—over what's yours and what's not.

To repeat, this is a rights issue. If enough people move to a DSNP-enabled environment and make a stink about the right to their personhood, lawmakers will act. But, remember, it's up to us to get the ball rolling. Government officials are followers, not leaders, in such situations.

Reaching Consensus

Aside from those cases when we'll have to involve traditional government, much of the work around governance, standards, and policies will spring from the self-governance approach described above. So, to round out that issue, we must also explore the mechanisms, such as voting, by which group decisions will get made. That is, how will we reach consensus? We must protect the new online communities formed through DSNP-enabled gatekeeping powers from the centralizing censorship of a single person, entity, or subgroup. How can group members make sure that curation and moderation reflect the wishes of the majority? How can users of an app gain assurance that the auditor charged with certifying its data policies is applying standards that meet their collective wishes and needs? In other words, how can this new system live up to the American Project's ideal of democratic self-governance?

Voting systems that would allow groups of people to express their preferences have struggled in the machine-centered internet, due to challenges around identity and privacy. This lack-of-identity problem was perhaps best summed up in another famous *New Yorker* cartoon, this one by Peter Steiner, who in 1993 drew a dog at a computer saying to another dog, "On the Internet, nobody knows you're a dog." Anonymity makes community governance vulnerable to so-called Sybil attacks—named for a 1973 book about a woman with dissociative identity disorder, then known as multiple personality disorder—in which a person or a company can create different logins to shift a group's consensus in their personal favor. Amazon and Yelp have long struggled with fake reviews and distorted ratings because people pose as different reviewers to submit large numbers of favorable reviews for their own business or unfavorable ones against a competitor. The problem is exacerbated by AI, which makes it trivial to spin up hundreds of thousands of fake identities. All of this is especially a problem for that cornerstone of democracy: the principle of one person, one vote.

Now, at last, solutions are at hand. Online voting will be more manageable in a world where we own our data and control our digital identities and where sophisticated record-keeping and privacy-protecting tools allow us to prove we are who we say we are without exposing sensitive personal information. Already, new forms of empowered decentralized communities use these tools to implement online voting in the management of shared resource decisions.

These advances could underpin a new era of self-governance for a human-centric internet. Groups of autonomous members could vote on acceptable content standards for their site or elect moderators to apply those standards. Or they could approve the deployment of new technologies that aid in that moderation effort. They could also use these decentralized voting systems for dispute resolution. If someone feels they've been unfairly censored, they could lodge a complaint and seek redress from the community, perhaps triggering a review and a subsequent ruling by a committee of elected delegates.

One particularly intriguing model for community governance is the idea of data cooperatives. This concept, promoted by MIT scientists and Project Liberty advisers Alex "Sandy" Pentland and Thomas Hardjono, is founded on the notion that if people can pool their data, they will be in a stronger bargaining position with any business that wants to buy it, much as unions strengthen workers' hands when it comes to wages. Pentland argues that these collectives should be formed around apolitical geographic units such as census blocks because it would drive them to coordinate data usage and content curation decisions around "what matters to them at the local level." Imagine if some of the most important uses for the internet revolved around intrinsically nondigital matters such as the local school board, the fire department, or your town's roads and infrastructure.

The discussion in this chapter is intended as a reflection on what's possible, not as a pitch for particular models which would violate the very principle of self-governance. What we are simply suggesting is an evolution of models and behaviors from the pre-internet age to the age we're living in. We will need to determine—through deliberation, collaboration, experimentation, and iteration—the path that suits us best. When approaches fail to serve our shared needs, members will seek out alternatives that do a better job at it. This power of the "exit"—which, of course, is only possible when we have full portability of the data and content that we own and control—should function as a disciplining force, encouraging communities and their leaders to respect the interests of the members. It's an accountability mechanism, one that will make manifest the "We the People" foundation of a fair, constructive, and inclusive communal life online.

~

In chapter 6, we put all of this into the context of the looming challenges and opportunities framed by the rapid onset of generative artificial intelligence, a technology that could dramatically transform our way of life. We will contemplate the better future that our proposed human-centric framework portends for the internet while also confronting the severe consequences of not embracing it.

CHAPTER SIX

Enter AI: It's Decision Time

In the opening chapter of his acclaimed book *Deep Medicine,* Eric Topol, a medical doctor, tells the story of a distraught mother bringing her newborn boy into an emergency room after he had started having repeated seizures. The eight-day-old infant was suffering from a condition known as status epilepticus. The prognosis, if a cure wasn't quickly found, was brain damage and, likely, death. CT scans showed a normal brain, with none of the electrical signals typically associated with epilepsy. A slew of potent drugs failed to stop the steadily worsening seizures. Doctors were at a loss.

As desperation set in for the boy's parents, a blood sample was sent to the Institute for Genomic Medicine at Rady Children's Hospital in San Diego, which carried out a rapid whole-genome sequencing. Some 125 gigabytes of data were processed, including 5 million locations where the boy's genome differed from the most common one. A natural language processing AI ingested the infant's medical records and sifted through those 5 million genetic variants to find 700,000 rare ones, of which 962 were known to cause diseases. Comparing that with the boy's phenotype data, the system identified a gene called ALDH7A1, which causes a metabolic defect leading to seizures. It also revealed that its debilitating effects can be overcome by reducing the intake of amino acids. The entire process took twenty seconds. The boy's diet was adjusted, his seizures ended, and he was discharged thirty-six hours later, after which he continued to live a healthy life.

As Topol's happy ending suggests, artificial intelligence has enormous potential for good. Sadly, we also know that AI can equally inflict tremendous harm. Scenarios are practically endless. Ransom-seeking criminals are using AI agents trained on the voices of children to give their parents the false impression they have been taken hostage. Political operatives are creating deepfakes to spread unfavorable impressions of their opponents, while terrorists are creating them to sow panic and unease in the public. Phishing attackers are finding that AI tools make it much easier to coax passwords out of unsuspecting computer users.

The enormous power of AI threatens to subjugate humanity to an even greater degree than the manipulative algorithms and other technologies of the preceding two decades. If we simply integrate AI into the same model with which the web has run, we are in grave danger. We would be taking that flawed model and, rather than first fixing it, putting it on steroids.

As we've noted, we stand at a fork in the road and must decide between two divergent paths. The decision we face is our starkest one ever. Down the first path is a new world of positive technological advances, one where humans control the data on which the AI models run, where the code they employ is fully transparent, and where those deploying the models have accountability. Down the second path is the same closed, machine-centric, and centralized data model that has underpinned the second generation of the internet. With all we know, with all we've discussed in this book, why would we allow a more powerful version of the same broken system to advance? Choose that second path, and we get dystopia, an even bigger retrograde step into the age of feudalism. Choose the first, and we can create the greatest civilization ever. This is not hyperbole.

The Real Threat: Badly Incentivized Humans

AI is not new. The two-decade-old social media, search, and shopping algorithms we've addressed in prior chapters are themselves forms of AI. The explosive new field of generative AI, where machine-learning algorithms deliver content and ideas that mimic human creativity, is just a more powerful version of technologies that Google, Meta, Amazon, Microsoft, and Apple have developed and employed over the past two decades. Both ingest large amounts of data so that they can make predictions about what humans want. The one big difference is that the datasets on which these so-called large language models (LLMs) are trained are many magnitudes more extensive than those used by social media algorithms. This means that, under the centralized corporate framework within which generative AI development is currently occurring, the only viable players are those that control vast amounts of human data and have the giant computing capacity needed to power these models. Who are those companies?

You guessed it: the same big data accumulators that have been surveilling us and building giant computing apparatuses all these years.

Of course, the products that LLMs deliver and their uses differ from those of social media apps, which changes the dynamics. Both are built on predictive, probabilistic models. But there's a particular recursive effect that's rapidly accelerating the development of generative AI. Since LLMs are essentially learning from their own output and how we respond to it, they are getting better and better at mimicking us as they go. Everything is intensifying, including the growth and power of these big platforms.

If Americans were to approach this technological leap in the same way they embraced transformative innovations in the past, they'd be giddy with excitement over this breakthrough. After all, the potential benefits lie not only in near-instantaneous medical diagnoses such as the one that saved that newborn infant's life. They're in climate science, energy efficiency, education, epidemiology, finance, sociology—pretty much any field of inquiry, endeavor, or industry. In all these areas, AI and machine learning are already boosting productivity and driving innovation. It's why many technologists and economists believe the global economy is on the cusp of secular change, with the cost of pretty much everything that is data-dependent poised to plunge.

But if you listened in the spring of 2023 to many AI scientists, the people you'd expect to be the most bullish, you'd have heard a decidedly gloomier story, a dystopian one even. Witnessing the launch of GPT-4, OpenAI's mind-blowing new version of the LLM that runs its generative AI products, many influential figures began warning that AI could evolve to turn against us. With more than 100 million users also having gleefully signed up for the same company's ChatGPT chatbot (built on GPT-3.5) to make it the fastest-adopted consumer application in history, there was a palpable sense that machine learning algorithms were experiencing an exponential advance in capability. For many of those steeped in the science of AI, it was a moment of "Whoa, hold on, let's slow this down!"

So, on March 22, 2023, just eight days after GPT-4's unveiling, a group of scientists, business leaders, and public figures, including Elon Musk, Apple cofounder Steve Wozniak, the historian Yuval Noah Harari, and Stuart Russell, a leader in the human-compatible AI movement, signed an open letter calling on "all AI labs to immediately pause for at least 6 months the training of AI systems more powerful than GPT-4." Citing the "profound risks to society and humanity," the signatories warned of an "out-of-control race" to deploy machine learning systems "that no one—not even their creators—can understand, predict, or reliably control." If the AI labs don't immediately pause such research, they said, governments should "institute a moratorium" until the risks and benefits of these systems are better understood. Six months later, more than thirty-three thousand people had signed the letter. No pause was instituted, however.

The letter was a lightning bolt. To many, it was a welcome reversal of the "move fast and break things" ethos that Silicon Valley had dangerously championed for years. But for Eliezer Yudkowsky, a pioneer in the field of artificial general intelligence, it didn't go far enough. Writing in *Time* magazine, he called for an indefinite worldwide halt to such research and said that, if necessary, the U.S. government should intervene with military action to enforce it—even at the risk of nuclear retaliation. Why was such drastic action necessary? "If somebody builds a too-powerful AI, under present conditions," Yudkowsky wrote, "I expect that every single member of the human species and all biological life on Earth dies shortly thereafter."

To be sure, Yudkowsky has long been seen as an outlier among thinkers on artificial intelligence, as someone with rather unorthodox ethical and practical views on how society should function. But although the bleak predictions in his *Time* editorial went beyond those of other critics, it was certainly a wake-up call, one that demanded we consider worst-case scenarios. As one observer put it to us, "Only human beings could be both intelligent and stupid enough to create their own apex predator."

But let's step back for a moment. If AI is to be safely developed, we need to be clear about one thing: The real threat to human existence does not, in and of itself, come from the machines. Yes, Hollywood has for years stoked our imagination with visions of robots turning against people. But that dystopian vision really stems from a foreboding fear of the true threat: the humans themselves. If AI mistreats us, it will be the outcome of design and deployment decisions made by us. I assume that most people working on AI have good intentions. Unfortunately, good intentions won't help us when an AI agent recursively develops autonomous capabilities and starts making its own decisions on how to fulfill its designated mission. If that mission is to maximize the profits of companies that already treat human beings as products, then those decisions will likely be detrimental to society. The best way to avoid the dystopian outcome flagged by the letter-writing scientists is to ensure that AI business models don't intrinsically create it. Let's use our "rewards" thinking to align the incentives so that, yes, the technology can advance but civilization doesn't pay the price. This will happen when the goals and interests of the companies building AI are aligned with society's. Here, the lessons learned from the current internet could not be clearer.

Will we continue the same extractive, exploitative business model in the AI era? If we do, the problems we've described in the preceding chapters will now go into overdrive. Human beings will forever be pawns in an even more powerful machine-driven system. This new era of digital feudalism will make the Middle Ages pale by comparison.

In the interests of weighing all options to combat that threat, let's put an extreme one on the table: We could, if we so choose, adopt China's solution to Big Tech power. There, the current internet's autocratic technology model and surveillance features sit comfortably within an autocratic political system—it's just that Chinese platforms' mind-manipulating algorithms serve the interests of the state, not private corporations. The Chinese government is, for example, using the powers of state to shape how TikTok influences impressionable young minds. It requires that children under the age of fourteen use Douyin, the original version of ByteDance's hugely successful app, in "teenage mode." That limits them to forty minutes a day, with an automatic cutoff at 10:00 P.M., and interrupts the kids' mindless scrolling with mandatory five-second delays. Children in China are also shown specially selected content addressing scientific, educational, and historical themes, along with a solid dose of Chinese patriotism and appeals for social cohesion. Meanwhile, China lets ByteDance deliver the no-restrictions version of TikTok to whomever wants it, of any age, in the United States and other democracies, where the app's addictive scroll function has preteens obsessed with mindless dances and facial expressions. You might say China is feeding its kids vitamins while serving sugar to ours.

To authoritarian-minded folks, perhaps this approach sounds appealing: Get a strongman dictator-type to issue a top-down mandate forcing the internet platforms to stop messing with young people's minds. Done. But, no, we live in a democracy for a reason. We cannot possibly allow such state control over our information system. That, for sure, would be the end of the American Project. So, instead of throwing the baby out with the bathwater, we, the citizens of democracies, need to take matters into our own hands. We must make the leap to a new system that's in keeping with the values of the kind of advanced society most of us say we believe in.

Of course, we want the U.S. government to back this effort and, as discussed, we will eventually need its help with some elements of the shift to a next-generation internet. But for the most part it is up to us, the citizens, to ensure that our operating system for the AI era—incorporating all the individual, societal, economic, and governance components that determine how we coexist—is compiled in such a way that it serves humanity's long-term interests. If we do this, we can foresee a deluge of positive stories about lives saved like those highlighted by Dr. Topol. AI could well become the most powerful driver of prosperity and well-being since the dawn of civilization.

It's worth noting that if we succeed in achieving this redesign, it would dash China's hopes for geopolitical supremacy over the United States. Beijing's surveillance model would be rejected by people elsewhere because the American Project–inspired alternative would be so much more appealing. America's secret sauce has always been its commitment to freedom. Now, if we develop and promote a human-forward model for the twenty-first-century internet, one that holds that same principle of liberty at its core, that projection of what Joseph Nye calls American "soft power" will be even more profound than it was in the twentieth century. China's autocratic model can't win against this, which means that Washington, not Beijing, will be in the driver's seat, determining the direction of the digital global economy for the next hundred years.

It would be madness for the United States to come at this from the other direction and try to beat China at its own game by developing a similar state-led model of digital autocracy. That then leaves us with a choice: Either we stay on the current path, that of a private, corporate autocracy, or we take a new path down which lies liberty, dignity, and the resurrection of the American Project. Which path will we choose?

Counterintuitive though it might sound, we believe following the lead of the Silicon Valley doomsayers who championed pausing the development of AI will take us down the wrong path if that measure is not also combined with moves to limit the power of the leading LLM providers. Unless it is accompanied by measures that strip them of their centralized control over our data and that require them to open-source their algorithms' codebases, any effort to halt research at the GPT-4 level would freeze development in favor of the same few companies that have presided over the internet's unhealthy evolution.

Here's a loose accounting of where things stood at the end of 2023:

- Over four years, Microsoft had reportedly invested more than $13 billion into OpenAI, much of that in the form of the necessary computing power that OpenAI lacked, all in return for 75 percent of the company's profits until it recoups its investment, while it was also aggressively developing an AI-powered version of its Bing search engine.

- Alphabet's Google had poured an estimated $200 billion into AI research over the past decade, the fruits of which were showing up in its Bard search engine.
- Meta was projected to have spent $33 billion on developing LLMs for generative AI in 2023 alone, with its existing Llama2 system about to be surpassed by one that would be twice as powerful.
- According to *The Verge*, Apple was investing millions of dollars per day in its model, Ajax, reportedly more powerful than ChatGPT.
- Amazon was said to be racing to build a version called Olympus to take on ChatGPT, Bard, and the others. It had also invested $4 billion in the OpenAI competitor Anthropic.

You get the picture. It's a tech arms race bigger than any we've ever seen.

This arms race has resulted in a massive concentration of computing power inside the tech giants' data farms and has given Nvidia, the biggest maker of the chips needed to run LLMs' vast data processing function, a trillion-dollar market cap. A government-mandated pause in AI development would amount to legally enshrining an AI oligopoly by barring competitors from developing superior alternative technologies—a convenient complement to these companies' existing internet oligopoly. We know where that market structure has gotten us.

Although we welcome the "Let's slow things down" spirit of the "Pause" letter, we also acknowledge that stopping the development of AI is exceedingly difficult, if not impossible. If U.S. or European tech companies complied with a moratorium order, how would their governments prevent Russia's and China's armies of data scientists from surreptitiously developing systems to weaponize against democracies? AI technology is going to advance whether we like it or not. Our most urgent mission is to ensure that it does so in a way that advances all of humanity's interests.

Let's level-set: To achieve a net positive outcome for society, it is not enough to give lip service to "human-friendly AI principles." We need to fundamentally change the information ecology within which this technology is emerging. Now, more than ever, we must put personal data under the control of the people who generate it, and we must embrace open-source code. In so doing, we will rightfully reclaim this technology as a collective human enterprise, rather than allowing it to become the next big thing for Big Tech.

Theft

The LLMs behind AI services such as ChatGPT and the image generator Midjourney—both of which require a paid subscription—are built on vast arrays of computing nodes coalesced into an artificial neural network (mimicking the structure of the human brain). To come up with all those impressive answers, these computing networks are "trained" on massive troves of information—many have already ingested more than a trillion separate items—from all the books, online posts, FAQs, speeches, videos, music clips, professional journals, and everything else that the open internet makes available. It is no exaggeration to say those services are monetizing the collective knowledge, ideas, and wisdom of billions of human beings over hundreds of years. These companies are profiting from our collective genius, something that truthfully should belong to all of us.

Yet, for all the amazing, natural-sounding answers these systems produce, an LLM is just an incredibly fast calculator. When ChatGPT starts answering a query, it is not so much "thinking" as it is running through gazillions of probability calculations. Driven by a giant mathematical function, it analyzes the query against the massive associations of words, phrases, and contexts it encountered during its training. It then starts choosing each successive word based on these probabilities, progressively crafting them into a phrase, then a sentence, then a paragraph, and so on. All along, it is driven to arrive at a combination that it estimates will most likely satisfy the request. This all happens in seconds. Often the results are impressive. But sometimes, these machines hallucinate. No one knows why.

If we recognize that this is just math, perhaps there's some hope that the creative human mind retains a dominion that machines can never occupy. Our ideas stem from an incredibly complex, highly evolved capacity to make unusual associations, to think in a combination of analytical and moral frameworks, and to draw on our emotional connections as well as our mutual, innate understanding of what makes us all tick. That's not the realm of a machine. Can a machine believe or disbelieve in God? Can a machine love? Can a machine have a sexual preference or, for that matter, any preference—for food, art, scenery? If a computer were programmed to kill someone, could we put it on trial for murder? Does an algorithm have morals?

We think not.

These are, by definition, elements of the human condition. Regardless of how far computer algorithms will develop, will they ever gain the kind of self-aware introspection that makes us humans unique? We don't think so. But that doesn't mean we shouldn't stay vigilant. Because, as we've seen from our internet experience over the last couple of decades, an algorithm can take away the free will and the authenticity that underpin human qualities even if it doesn't acquire them itself. Whatever we do in terms of managing our relationship with AI, it must protect and elevate these defining core aspects of our humanity.

Certain AI systems are already outperforming some humans in some measures of creativity. In one example, researchers asked different GPT-based LLM services to come up with as many uses for a rope, a box, a pencil, and a candle as possible. Their responses, alongside those of more than two hundred humans, were then assessed for originality and creativity in an anonymous test by independent evaluators. On average, the chatbots outperformed the human group, though the highest-scoring results were still from humans. That might sound like we're ultimately doomed as AI technology inexorably advances, but two factors greatly downplay the significance of this result. The first is that it's hard to distinguish what's creative and what's simply a machine tapping a list of options that's naturally larger than a human memory can conjure. The second is that it is entirely possible, in fact quite likely, that included among the massive amount of internet-based information on which OpenAI's LLMs have been trained are a series of these types of tests, known as "alternate uses tasks." So, the LLMs were not so much being creative as regurgitating past test results. You might say they were cheating.

The second point is extremely relevant to our case. OpenAI's system is a black box. We do not know what data has been fed into it. The word "Open" in the company's name is a glaring misnomer. OpenAI, Inc., was set up in 2015 as a nonprofit with an open-source codebase and a philosophy of shared research, partly at the behest of one of the founding board members, Elon Musk, who expressed concerns that AI could do great harm to humanity and has since had a bitter falling-out with CEO Sam Altman. In a blog post at the time, the company said, "Since our research is free from financial obligations, we can better focus on a positive human impact." But four years later, it did an about-face and went all in on financial obligations. A for-profit subsidiary known as OpenAI Global LLC was created so that it could attract venture capital and offer its employees shares. That for-profit subsidiary's code is now closed-source. And with Microsoft's whopping investment in 2023, it is now also phenomenally well capitalized. The officially nonprofit OpenAI, Inc. continues to be a 2 percent shareholder (Microsoft owns 49 percent and employees and other stakeholders own the remaining 49 percent), and its board has nominal control over the for-profit entity under its unorthodox governance structure. However, following the short-lived leadership drama of November 2023, when Altman was fired and rehired in the space of just four days, the power of the nonprofit board looked greatly depleted. A backlash over Altman's ouster among OpenAI employees—many holders of the for-profit company's now valuable stock—and a resignation by Altman's cofounder Greg Brockman led the board to hastily reinstate Altman and to reconstitute its membership. Pivotal in all this was a sudden offer that Microsoft made to Altman and Brockman to form a new lab inside the tech giant.

Following the failed leadership shuffle, three of the board members who'd voted to fire Altman were fired, among them the AI researcher Helen Toner, who had reportedly argued with the CEO over an academic paper she'd written with two Georgetown University colleagues. Among other things, that paper criticized the premature timing of the company's ChatGPT release. Reuters also reported that some researchers at the company had gone to the board with their concerns about a breakthrough discovery that they feared could threaten humanity. It seems these developments, coupled with concerns about management accountability, led the board to make its drastic move, one that ultimately backfired.

Perhaps the sudden, ham-fisted way in which the OpenAI board handled this crisis was ill-advised. There may have been a smoother way to address their human-safety concerns. And it now seems unwise to have concentrated the oversight of such important technology with just a handful of people, when a truly open-source model would have allowed much greater oversight with a significantly larger group of people reviewing the company's work. Nonetheless, we find it exceedingly disappointing that this affair was often portrayed in the mainstream press as a failed "coup." As we read it, what happened is that Altman's reinstatement essentially stripped the one institution charged with keeping him and his fellow executives accountable for human welfare of its watchdog capabilities.

Upon his return, Altman agreed that the board could conduct an internal investigation into his conduct, but the entire affair has made it clear that commercial interests have won this round. We don't know whether the nonprofit board ever had visibility into OpenAI Global's black box or into the commercial deals that it cut—OpenAI's governance model is almost as opaque as its algorithms. One thing's for sure, though: None of us have any way of knowing how OpenAI's machine is using our data and developing its output. We're stuck with the same business model that dictates how Google uses its search algorithm and how both Zuckerberg's Meta and Musk's X deploy their curation algorithms.

We find the outcome of this bizarre turn of events quite ironic. If you'd heard Altman speaking as recently as 2018, you would have assumed that he would be firmly on the side of Helen Toner and the others who were advising extreme caution about the rollout of OpenAI's technology. In raw footage I was shown from a never-released documentary about artificial intelligence, Altman was asked what was at stake in all of this. His answer: "Like the next two billion years of the universe. This is, you know, either we get this wrong and the light of consciousness kind of goes out of the universe as far as we know, or we get this right and the children of humanity go off and expand and colonize the universe. And that's like very high stakes, right? Like either this is an end or a beginning."

It's worth contrasting this victory of no-holds-barred capitalism with China's approach, though of course neither is appealing. In China, the executives at ByteDance, Tencent, Weibo, and other powerful internet companies will act at the command of the government. (Some U.S. commentators have accused Beijing of directing the explosion of anti-Semitic content seen on TikTok during the Gaza conflict.) In the United States, by contrast, we've so protected the Big Tech platforms from any real oversight—

even when their monopolistic systems run counter to antitrust and other laws—that we're powerless to assert society's interest over their actions. It's time for people power. We can't trust these companies and their powerful machines to look after us.

Google once championed its motto of "Don't be evil." But the company dropped it in 2015, a reminder that slogans are just that. The incentives baked into a company's business model, not lip service, ultimately dictate its actions. Similarly, don't be fooled by any of the compassionate-sounding words that sprang from the mouth of Altman during his tour of Washington promoting a captive regulatory approach. With him and all the other Big Tech CEOs calling for rules that would protect their privileged position from unwanted competition, we can be sure of only one thing: As long as they own and control our data, they will be incentivized to make money and to gain market share and influence by exploiting it. This will remain the case until the underlying model changes.

Meanwhile, artists, filmmakers, musicians, and writers are finding LLMs indiscriminately scraping their creations from the internet, inviting copycats into a free-for-all. It's outright theft. The digital artist Greg Rutkowski told the BBC that his name had been used in prompts for AI image-generation tools more than 400,000 times in the ten months following September 2022. Type the prompt "battle scene in the style of Greg Rutkowski" into Midjourney, and the image generator will produce a striking picture in a style indistinguishable from that of the digital artist. Lawyers are already going to bat for film studios, publishers, record labels, artists, musicians, collectors, and others who are claiming that AI systems are ripping them off. And, in an encouraging, prohuman decision, the UK's highest court ruled in December 2023 that inventions created by an AI model are ineligible for patents because "an inventor must be a natural person." As of this writing, the U.S. Copyright Office has ruled that art generated solely by AI is not eligible for copyright protection. But the bigger question around the rights of artists whose work is ingested into an LLM remains unresolved as a pivotal U.S. Supreme Court ruling looms.

The theft we're concerned with goes beyond the works of noted, copyright-registered artists. It's the appropriation of the work that everyone—humanity at large—has done over millennia. The massive data trove that LLMs are accessing, courtesy of an internet model that never granted control over their data to the people who create it, is a pretty good proxy for the sum total of human knowledge and creativity since the dawn of civilization. Under U.S. law, copyright lasts either the life of the author plus fifty years, or seventy-five years from publication, or one hundred years after creation, whichever is shorter. So, the classics, the works of philosophers such as Thomas Paine, are fair game for everyone, including the LLMs. But the reality is that all of that seventy-five-year-old, public-domain material is minuscule compared with the amount of content and data we humans are now creating on a daily basis through the lives we lead. Per a famous remark in 2010 from Google's then CEO, Eric Schmidt, humanity was at that time generating five exabytes of information every two days, the equivalent of all that had been "created from the dawn of civilization through 2003." Now, in 2023, it is estimated that computers are creating two-thirds of a zettabyte of information every two days, or six hundred times the amount in 2010. Think of that mountain of information as *our* work, because it is. That's what's at stake. Of course, a lot of that is user-generated content on social media sites whose terms and conditions strip people of their right to enforce such claims. But consider the dynamic on a deeper level: A big chunk of the data that these machines ingest and then turn into a product is the sum total of your conversations, social connections, and interactions. It is inherently a reflection of your human essence. For the purposes of the digital economy, it *is* you. How is it right that the life you've lived is being vacuumed up into a commercially owned black box to do with as it will? Now that you're six chapters into this book, you'll know where we're going with this: That data and content should be yours as a right, regardless of what the Big Tech platforms' terms and conditions say. Those systems can absorb that information, monetize it, use it against you, and even weaponize it. It is an egregious breach of human rights.

That stolen material is now, dangerously, being repackaged as disinformation. In a report about how AI moviemaking tools spawned a plague of fake science videos, a BBC film crew showed a group of schoolchildren a clip in which the narrator claimed that the Egyptian pyramids were built by aliens to transmit electricity. In interviews, kids routinely said they believed the account and were impressed by how convincing it was. There was such a proliferation of these videos that YouTube's suggestion algorithm quickly surfaced them as the kids searched science-themed topics. How-to videos on YouTube explain how to make the most of the AI tools that produce these videos and how to maximize ad revenue—50 percent of which is taken by YouTube's owner, Google.

Meanwhile, deepfakes are interrupting public discourse and the law. Lawyers for Elon Musk—already the subject of numerous deepfakes, including one that falsely showed him high as a kite on drugs—claimed that recordings of their client making statements about Tesla cars' self-driving capabilities should not be admitted as evidence into a case surrounding a fatal accident because those recordings could have been deepfakes. Here was the world's richest person looking for a payout of the liar's dividend. In that case, the judge rejected the lawyers' argument and called it "deeply troubling," but no one believes that discerning the truth is going to get any easier.

To top it off, there's the threat to regular work. Here's how that same omnipresent man, Elon Musk, put it in a November 2023 interview with British prime minister Rishi Sunak: "There will come a point where no job is needed. You can have a job if you want to have a job for personal satisfaction. But the AI will be able to do everything." To his credit, Musk didn't just regurgitate the upbeat readings of some in Silicon Valley, who celebrate the idea that humans will be free from the drudgery of work and that universal basic income (UBI) distributed by a government, or perhaps a private entity like Altman's Worldcoin project, will allow people to enjoy life's pleasures. Rounding out his thoughts on the job loss problem, Musk concluded by saying, "One of the challenges in the future will be, how do we find meaning in life?"

It's more than a "challenge," though. It's an existential threat. Not because we should glorify work and want everyone to toil in factories or dull office jobs, but because, as we discussed in chapter 2, the prospect for self-actualization is essential to the human condition. We need to elevate the individual to their rightful place. What purpose do we otherwise bring to our life? The answer to that could come from work, or maybe it will come from art, but it won't be satisfied in a positive manner if the individual isn't in charge of their life. UBI or no UBI, we cannot be reduced to merely being the quarries from which digital machines mine the information needed to enrich their owners. There must be agency in the AI era. We must control the data.

Truly *Open* AI for All Humanity

It doesn't have to be this way.

If we can gain a say over the makeup of the all-important AI training datasets and real visibility into how they are used, we can keep those who manage them accountable to shared human and planetary interests. If we control the data at its source, we can start to rely on smart, open-source software coders to build trusted monitoring systems that employ unique digital identifiers and distributed-ledger public record-keeping methods to tag and track the AI systems' inputs. That way we can verify that they come from trustworthy, legally sound, human-friendly sources. It will help us not only contain copyright abuse and more nefarious uses but also, eventually, enable a fair, humanity-affirming way of compensating people whose data and creative work is used to power these machines.

It's not just about stopping bad actors from taking intentionally harmful actions. It's about constructively involving everyone in the discussions, debates, and work needed to overcome antisocial outcomes. There should be open discussion and debate over AI training models and how they should best be designed to overcome their embedded racial, gender, or political biases. (We see this in early facial recognition models, trained on white faces, that don't recognize Black people, or in the way that AI-generated images overly sexualize women.) Having societal input into these processes is vital, but it's only possible with an open-source model, where third-party programmers can monitor the data inputs and suggest updates and improvements to the code.

Silicon Valley investors will inevitably push back against opening their proprietary software for all to use and copy. But builders of these systems will still have plenty of ways to profit. There will be opportunities to develop pay-for-service apps and interesting new ad-generated business models on top of open-source LLMs. (Want an AI-based assistant to help you with your daily exercise routine? There'll be an app for that!) Much like our vision for the internet economy, the AI economy must treat its all-important, foundational knowledge systems as part of the public commons and individuals' data as theirs. And no more black boxes. We citizens deserve to know what's going on with systems that will have such a transformative impact on our lives.

As discussed in chapter 5's treatment of app audits and standards, the adoption of decentralized digital identifiers and open social graph data will empower this kind of governance. This will make for a safer, more trustworthy world, one in which truth can win out. With the same tagging and tracking systems, we can prove, for example, that a video was sourced from real live-action footage and was not modified by a deepfake generator. Imagine a check mark that verifies content wasn't AI-generated, something like the little lock signal that browsers now put before a URL to indicate that a website has a secure sockets layer (SSL) certification and that it is safe to use. We can build the same kinds of protections, models that don't censor material but inform users so they can make educated choices, into the next-generation internet. But only if we all control our data and how it is used.

Such adjustments are vital if we are to save our political process. Without checks and balances against machine-created disinformation, one can easily imagine, say, Russian or North Korean agents using fake but highly convincing videos to peddle conspiracy theories and incite vigilante mobs into violence that tears the country apart. The result is a vicious cycle of mistrust, misinformation, and social breakdown breeding ever more of the same. If we can neutralize these threats, however, and start to revive trust in information, we can restore the kind of positive feedback loop that the American Project enjoyed, bolstering a sense of ownership, agency, citizenship, and shared economic opportunity in the AI age.

The Bright Side

Fix the downside risks, and we can start to embrace an optimistic outlook on the huge potential these digital technologies hold. We should strive for the kind of energetic excitement that John F. Kennedy generated about going to the moon—perhaps even more so because these are innovations that we've all collectively contributed to—rather than the fear that's currently dominating the discourse around technology. As Franklin D. Roosevelt famously warned, fear is not something to be embraced; it is, rather, "the only thing we have to fear." Negotiating and setting policy in a state of fear is a recipe for mistakes—such as agreeing to a pause in AI development that secures the monopolies of the front-runners.

Still, given the technology malaise prevalent in the current era of the internet, it's perhaps understandable that people fear AI. We saw this play out in the 2023 strike by Hollywood writers and actors. The original culprit behind the inadequate pay that prompted both groups to halt work was the business model of centralized streaming platforms such as Netflix, which had made the economics of anything other than blockbuster moviemaking unsustainable. But what pushed the actors over the edge and into joining the writers on strike was a proposal by the Alliance of Motion Picture and Television Producers for studios to use background actors' likenesses in AI-generated scenes. Angry actors began comparing the annual salaries of less than $26,000 that 87 percent of them earned with streaming platforms' hiring campaigns for AI experts, with Netflix posting one such position at $900,000 a year. Quoted in *The Intercept,* the actor Rob Delaney nailed the perfect response: "AI isn't bad, it's just that the workers (me) need to own and control the means of production! . . . My melodious voice? My broad shoulders and dancer's undulating buttocks? I decide how those are used! Not a board of VC angel investor scumbags meeting in a Sun Valley conference room between niacin IV cocktails or whatever they do." The idea of "$900k/yr per soldier in their godless AI army when that amount of earnings could qualify thirty-five actors and their families for SAG-AFTRA health insurance is just ghoulish," Delaney said, adding, "Having been poor and rich in this business, I can assure you there's enough money to go around; it's just about priorities."

Priorities. That's what all of this is about. What do *we* as a society want from technologies like AI and the internet? What approaches and models will serve the broadest public interest? We must put ourselves in a position to voice what we want, to collectively define the priorities for all of society. If we can do that, as Delaney said, we'll find that "AI isn't bad." In fact, we can steer it into delivering outcomes that are hugely beneficial to humankind and the planet.

This is an exciting opportunity, a once-in-a-civilization opportunity. If we can forge a pro-human, prosocial, pro-planet framework for AI development, we can collectively steer this unique, computational extension of our shared human knowledge toward solving the many problems we face. Whether it's to improve weather prediction and climate modeling; to help cities optimize power grids and factories to craft efficient energy consumption strategies; to fast-track disease diagnosis, epidemic mapping, or drug discovery; or to help students absorb information more quickly so they can focus on being better creators and strategists, AI can change the world.

We have the contours of what a new internet would look like. We just need to claim ownership of it. We need to get our minds into that elevated place where change can happen—much like all those who dreamed of a better existence in the eighteenth century, the ordinary people who decided enough was enough with the broken system they'd toiled under and then set out to build a new American civilization. If we can act with the same spirit and resolve, we can restore human agency and start making something out of the digital lives we are living. If we do that, the future for all of us, for the planet, is bright.

CONCLUSION

Ctrl-Alt-Del-Esc: The Future Is in Your Hands (Actually, It's at Your Fingertips)

Imagine you're living on a farm in the west of Ireland in 1850. The potato famine—or what you call the Great Hunger—has been raging for five years. Despite the pleas of desperate Irish advocates, the British Parliament, in which you have no representation, has repeatedly refused to mandate a halt in food exports from Ireland, which is contributing to widespread shortages. Close to 100,000 of your fellow citizens have died of starvation, including one of your daughters.

Then a cousin comes to you and says he's heard there's a new life to be had across the Atlantic, in America. You could get your own plot of land, one that *you* would own. It would be a chance to break your dependence on that absentee landlord who lives off your rent payments in London and does nothing to help maintain the land you toil. Or, your cousin suggests, you could start your own business in Boston, or New York, or Baltimore, or Philadelphia. American towns and cities are swelling with new arrivals, which means they are building out infrastructure and need laborers and tradespeople. Their residents are also demanding basic services for which they will pay a fair price. The United States' surging economy is the antithesis of this wasteland of death in which you are trapped.

You are sold on the idea. You and your cousin round up family members and friends. You scrounge up the fare needed to get all of you on a ship departing from Cork. You set sail for New York.

The circumstances that most Americans and others in the developed world face in the 2020s are not nearly as graphic as those that Irish emigrants faced in the mid-nineteenth century. Also, alongside the problems we have identified, the internet has brought some real benefits to our lives. Most Americans are not starving. And, despite all the failings recounted in this book, our democracy, more or less, continues to afford us citizens representation—at least for now. Still, as we've argued, our information system is severely broken. We think it's time, similarly, to "migrate"—not so much to a new land, but to a new internet.

What we are asking of you is far less painful than the choice my ancestors faced. The overcrowded, poorly built, disease-filled vessels that sailed from Dublin, Galway, Limerick, and Cork in the 1850s were called "coffin ships" for a reason: Thousands died in the dark holds during the Atlantic crossing. In this case, all we're asking of you is to make a choice: Maybe download a few apps and sign up for a service built on DSNP or some other protocol that gives you control over your social graph. There should be no bloodshed in this revolution.

Still, we think the story of Irish migration to the United States, a journey undertaken by an estimated two million people between 1845 and 1895, is a great metaphor for the opportunities that lie ahead if we all move to a more human-friendly internet. The dream of a new life in America is often cited for having kept up the spirits of those desperate souls crammed into the holds of the coffin ships. Similar stories of hope have also accompanied millions of people who, more recently, have left their homes in Venezuela, El Salvador, Honduras, and other downtrodden places to make the long, dangerous journey to the U.S.-Mexico border. They come despite their vulnerability to human traffickers and others who exploit them, despite the physical dangers lurking in the jungles and deserts they must cross, and despite the very high risk that they will be detained and sent back by U.S. Immigration and Customs Enforcement agents. Wherever you stand on the contentious issue of undocumented immigrants, let's acknowledge that the vast majority are drawn by a very powerful idea, that America, despite everything, represents the dream of the possible. We think it's now time for all of us to start to dream of something better for the next generation of the internet. So, let's indulge ourselves in that dream.

Before we do, an important caveat: As with the nineteenth-century United States to which my ancestors migrated, where half of the nation continued to support the heinous slave trade, the new internet will not be an unblemished utopia. Bad actors will still do bad things. There will be scams and manipulative conspiracy theories and stupid memes and toxic behavior. Sadly, humans gonna human. What's different is that this new internet will not be structured in a way that incentivizes bad behavior. Since there's no longer a clique of uber-platforms deploying algorithms whose reward system encourages extremist views or seeks to shape our behavior in their profit interest, space will open up for moderate voices and decent behavior. Improving society and addressing its many on- and offline challenges will be a work in progress. What this new framework gives us is a chance to make a fresh start. As with tens of millions of past and present immigrants to the United States, that's all we are asking for.

Our Internet

Welcome to the NewNet.

(This is simply our first stab at a name for the third-generation, people-centric internet, the successor to the second-generation version, the internet of data, which followed the first version, the internet of machines. We welcome other suggestions but feel strongly that naming the new internet is an important first step for all of us to stake a collective claim of ownership over it.)

To start with, here are some defining features of the NewNet:

Personal digital identity: You are recognized as a person, not a machine with an IP address, when you log on to the internet. You have one login, not a mountain of vulnerable passwords.

Choice: You decide what goes into your or your children's news feeds. You, not the platforms, choose what you read, view, or listen to. And, without the monopolies and walled gardens, you have an abundance of services and apps to choose from.

Digital property rights: You own whatever digital content you create, you own your social graph data, and you own the right to monetize it.

Autonomous permission: You decide who sees your content and connects with you. You decide if and when to revoke that access.

Opt in, not opt out: You choose when to signal your interest in something. You don't have to tell an application or platform not to surreptitiously track you with cookies and other surveillance tools that target you.

Portability: You can move your social graph data—your contacts, playlists, friends, followers, and so forth—wherever you like, from one application or platform to another in a form that is operational.

Individual terms and conditions: Rather than clicking on the terms and conditions of use pushed to you by a few huge platforms, you set the rules by which your data can be used.

Decentralized storage: The information of the internet is no longer stored in data centers monopolized by a few mega-companies. It is distributed across a vast array of autonomous computers with no central entity in charge.

A public network effect: Since the aggregate of our combined social graph data can no longer be monopolized by one company, the network effect that delivers value solely to the biggest actors is no longer theirs. The network effect is now a public good. Any innovative application builder or service provider can tap into it with permission.

Interoperability: Apps, platforms, and AI agents all talk to one another within compatible computing languages and cross-platform assets. There are no more walled gardens.

Authenticity: Just as there are consequences if you show up in the real world as a fake person, there are now consequences if you show up on the internet as a fake person. (This includes consequences for bots that create fake personas.) Your rights to keep your data private and to use pseudonymous avatars and identifiers are still protected as necessary to keep you safe, but lying about who you are is no longer a profitable undertaking unto itself. We all benefit from an internet made of people, not fakes and phonies.

These features have been made possible thanks to a re-architecting of the current internet's structure, now improved with DSNP. The core values of a free, democratic society are now embedded into it. Inspired by the American Project undertaken by the founding fathers, the NewNet offers citizens the best opportunity to pursue happiness, to experience prosperity, and to enjoy optimal physical and mental health. This amended architecture assures and encourages the rights, responsibilities, rewards, and rules that support the interests of individuals, society, a market economy, and our system of governance. It is the framework of a constructive, positive, evolving, iteratively improving system. With artificial intelligence now dominating our lives, pushing us to a point where our physical and digital existences are entirely inseparable, this structure is protecting our agency, dignity, and independence—our personhood.

In the NewNet, people are valued for bringing to the negotiating table the thing that matters most: their authentic selves, now backed by digital data about their lives and connections, data they own and control. Self-sovereign DSNP accounts have, for the first time, given them autonomy over who gets to see their information and data. And as human beings employ this to reclaim a position of dignity and authenticity around who they are offline and online, a state of mutual respect has reemerged. Division for division's sake and toxicity are now frowned upon. And our political process finds more space for moderate views as we arrive at an understanding of a common truth through a rational process of debate and deliberation rather than an endless, unresolved volley of point and counterpoint.

In the NewNet, the media industry, as vital as ever to a functioning democracy, is better equipped to deliver important, well-reported, balanced, and accurate stories that help us make decisions on the pressing matters of our day. News outlets are no longer algorithmically driven by a race-to-the-bottom clickbait battle. Without a data-monopolizing platform sitting in the middle, publishers get a clearer picture of their audiences, connect directly with their readers and viewers, and create sustainable business models that aren't dictated by mindless competition for attention. Truth, once again, is valued.

In the NewNet, cybercrime is no longer one of the biggest industries on the internet. It's not that rogue actors have given up on a life of crime or that software no longer has vulnerabilities.

What's happened is that we've made the economics of information and identity theft far less appealing. Since the data resides with us, it now costs too much and pays off too little for hackers to launch separate attacks against each node within a population of billions. It's just not worth the effort to obtain small caches of data from each one of us. And much of the most valuable data is now in the public domain, so there's no competitive advantage in stealing it. The other payoff from this is that all the engineering brainpower that previously went into building firewalls and other security protections is now being steered toward more proactive, positive projects that advance the human condition.

In fact, far beyond the work of those engineers and security experts, untapped creative and analytical human energy is being unleashed everywhere, which means we are undergoing an economic, cultural, and social renaissance touching every part of our lives. Now that individuals and businesses are freed from the intermediation of the platforms, the market for ideas has become inherently more vibrant. There is now unprecedented competition in many areas of the digital economy—which is to say the entire economy, since we are now well and truly in the AI age, from which no industry is isolated. Combined with the fact that much of the data—the lifeblood of this economy—cannot be captured by network-effect-seeking monopolies, this competition is fueling a positive feedback loop of innovation. Businesses that have tapped into parts of that data, all with the permission of its owners, are striving to outdo one another with real, value-adding products and services. New ideas and solutions are breeding other new ideas and solutions that are built on top of them, all of it supercharged by machine-learning models that are structured to serve humanity's interests.

These technologies, underpinned by a sustainable architecture that centers human flourishing, are beginning to solve some of the world's biggest problems. We have finally moved beyond the mistrustful bickering at the United Nations over which country owes what to whom for putting the environment in grave danger. Instead, we are taking real, mass-scale action with renewable energy and carbon-capture technologies to contain climate change, together. A new food revolution is also underway as smart, highly efficient solutions are more rapidly bubbling up for industries such as vertical farming and genetic protein manufacturing, producing an abundance of sustenance for the world. Homes are being built at a fraction of their previous cost, lowering the barriers of entry for millions. And the poorest of all are experiencing economic advances and enjoying, for the first time, real self-determination as new sources of energy and water combine with low-cost digital finance and management tools to empower communities the world over.

We see it, too, in the realm of the arts, a digital renaissance, with a proliferation of novel artistic forms. Much like that which began in the fourteenth century, the current renaissance is starting to capture the hopes and dreams of our age, a new creative expression of the human condition. With the once all-powerful platforms no longer acting as the arbiters of what gets seen or read, the aesthetic of the NewNet is vastly different from the previous era. Creators of articles, books, music, photography, film, and fine art are no longer separated from their fans, audiences, and patrons by data platforms, which in turn means the studios, galleries, record labels, and others that had previously acquired that data are now disempowered. (In one manifestation of this, going on strike is an irrelevant option for Hollywood's actors and writers since they are no longer dependent on giant production and distribution companies to reach audiences in the first place. It also means the "blockbuster or nothing" model is dead, opening endless new opportunities for indie ideas to thrive.)

Society is healthier, quite literally. Yes, there are still many people living with mental illness, but the system isn't incentivized for dopamine-triggering content that feeds on their vulnerability and pushes them over the edge. One hopeful statistic confirms this: Teen suicides are less common. Meanwhile, hospitals, universities, small pharmaceutical labs, and biogenetic start-ups are bringing forth an explosion of medical breakthroughs. Since the world's sensitive genetic and medical records are no longer under the control of insurance companies and health industry players that previously guarded them as a proprietary advantage, a flood of valuable information has supercharged the research process. En masse, people are now consenting to feed their genetic and medical data—all of it anonymized and protected with strong encryption—into collective efforts to find cures, treatments, vaccines, and other preventive measures for cancer, Alzheimer's, Parkinson's, COVID, you name it.

Most important, each of us *feels* better about our lives within this new system. Technology is no longer a soul-sucking, dehumanizing burden for us to bear. Instead, we are restoring our belief in its potential to improve our existence. There is a new sense of purpose, of possibility, of agency. Artificial intelligence and other fast-moving technologies are no longer seen as something to be feared but as tools we can collectively harness to solve problems together. There is excitement about what lies ahead. And while we still face the many challenges associated with coexisting on this fragile planet, we now once again have that all-important ingredient for a meaningful life: hope.

~

If this is a world you want, then what can you do to make it a reality?

Well, some steps you can take on your own. With the ease of your smartphone, laptop, or PC, explore self-sovereign identity solutions. You can download a self-sovereign wallet to help you engage with new applications based on decentralized data standards that adopt DSNP. Seek out AI providers building on transparent, open-source, and decentralized data processing models. When you do, say goodbye to Facebook, Twitter, Instagram, et al. You won't be needing them anymore. And if you want to learn more or if, perhaps, you have coding knowledge or other programming skills and would like to contribute to the technology, or simply wish to follow the progress as the NewNet is built, we suggest you check out ourbiggestfight.com.

The real change comes when we all act together. We want everyone to empower themselves and be an agent of it. Get the word out. Spark conversation and debate. Urge your friends and contacts to join you in the migration to the NewNet. Form new groups with them, employing whatever self-governance models you all feel are best. And in your workplace, where employees have long been important drivers of policy changes, make the case to management for embracing new business models that will thrive in an open-data economy.

Spread the word: There's a NewNet.

From David Quirke, a creative strategist and Irish raconteur, we heard a great suggestion for describing these action items in computing terms familiar to any PC user: Ctrl, Alt, Del, Esc. Take *control* of your data. Migrate to an *alternative* model. *Delete* the old one. And *escape* to a better web, for a better world.

It's your choice.

Acknowledgments

Frank

To my parents, who taught me about the importance of character, and how to fix problems rather than complain about them. And to my late brother, Richie, who was one of my heroes, and my other five siblings, who make me proud.

To all my colleagues at McCourt Global, who show up with their best every day and who because of their hard work and dedication have given me the gift of time, which has allowed me to write this book and focus on Project Liberty. A special thanks to my senior advisers at the company, and my friends, Jeff Ingram and Barry Cohen. It was Barry's idea that I write this book.

To Mark, Sophie, Ellen, Dan, Sara, Rick, Margarita, and all my friends who have generously supported me and this work.

To Jack DeGioia, Dan Porterfield, Angela Glover Blackwell, Dan Doctoroff, and Darren Walker, who were the original members of the Unfinished Council, the launchpad for Project Liberty.

To Maria Cancian and everybody at the McCourt School of

Public Policy, whose service and commitment to others and to the common good is an inspiration.

To everyone at Project Liberty and the visionary technologists on the labs team, all of whom have been quietly doing the hard work for the last four years.

To my coauthor, Michael Casey, who went above and beyond to help me produce this book. More importantly, I've found a friend.

To my seven children who constantly remind me whom, and what, I'm fighting for.

To the person who rekindled my spirit and fills me with joy each and every day, my wife and my partner, Monica.

Michael

I want to thank Kevin Worth and every member of the incomparable CoinDesk team. It was a highly rewarding, yet extremely difficult year for CoinDesk. Led by editor in chief Kevin Reynolds' dogged, dedicated newsroom and executive producer Joanne Po's multimedia team, we won the prestigious Loeb and Polk awards for our world-beating coverage of the FTX drama. Meanwhile, the company faced down a gut-wrenching contraction in the industry we cover and lived through an exhausting, drawn-out corporate acquisition. Throughout, my colleagues never ceased to support and encourage me as I took on this book project alongside my day job.

I'd also like to thank my literary agent, Gillian MacKenzie. Over a partnership of seventeen years, she has been a tireless, creative, and relentless supporter of my work.

In undertaking this project and other previous endeavors, I've been fortunate enough to have a group of people close to me whose advice and friendship have been vital, especially at the most challenging times. I can't name anywhere near all of them, but I'd like to single out Phillip Chambers, Lance Koonce, Jenna Pilgrim, and my podcast partner, Sheila Warren.

To Frank McCourt, thank you for your deep passion for and commitment to this vital issue, for being an exacting yet accommodating coauthor, and for encouraging me to join you in what is undoubtedly one of the most important projects of my life. The feeling is mutual, my friend.

Finally, there is no way I would have gotten through this year without the boundless love and support of my family. To Mum, Dad, and my three sisters, as far away as you are, you help me to keep going. And, of course, to my darling wife, Alicia, and daughters, Zoe and Analia, I'm eternally grateful. I love you more than anything.

The Authors

This book was brought together on a highly expedited schedule. To meet it required the assistance of a mini army. There are far too many soldiers to name here. But we are deeply grateful to every one of them who helped us get this project over the finish line.

We're especially grateful to David Drake and the team at Crown, who accommodated our tight schedules and unorthodox publishing ideas. Paul Whitlatch proved to be an incredibly efficient and clear-thinking editor. We are particularly thankful for the way David and Paul helped us evolve the concept for this book away from a longer format to the shorter, punchier manifesto-like work that we ultimately produced.

We'd both very much like to thank the McCourt "brains trust" of Paula Recart, Max Jenkins, John Loeffler, Liya Lehtman, Michael Drennan, Tomicah Tillemann, and Brin Frazier, who tested us and kept us true to the mission. Also, much gratitude to Shéhérazade Semsar-de Boisséson, who has played a variety of roles to help us on this journey. We were also blessed with invaluable advice on an earlier draft of the manuscript from Melanie Mattison, whose teacher's instincts helped us find language that would reach a broader audience.

We're grateful to various members of the extended Project Liberty community who weighed in with ideas and feedback, people like Frances Haugen, Braxton Woodham, Harry Evans, Constance de Leusse, Maria Ressa, Piedad Rivadeneira, Andrew Mangino, Hans Ulrich Obrist, Andras Szanto, Larry Lessig, Jonathan Zittrain, Maggie Little, Sergei Guriev, Martina Larkin, Jonathan Haidt, Beeban Kidron, Miguel Silva, Rob Reich, Stephen Gilbert, Nate Persily, Marietje Schaake, Sandy Pentland, Deb Roy, Simon Johnson, Erik Brynjolfsson, and Daron Acemoglu, to name just a few. And a special callout to David Clark, one of the early inventors of the internet, whose judicious assessment of this manuscript was invaluable.

ABOUT THE AUTHORS

Frank H. McCourt, Jr., is the executive chairman of McCourt Global, a private family enterprise working across the real estate, sports, technology, media, and capital investment industries. He is the founder and executive chairman of Project Liberty, a broad-based effort to build a better web for a better world. The project includes the development of an open internet protocol (the Decentralized Social Networking Protocol), which shifts data rights from platforms to people, and an institute—launched with founding partners Georgetown University, Stanford University, and Sciences Po—to advance research and develop a governance framework for the internet's next era.

Michael Casey is chief content officer at the award-winning media outlet CoinDesk, a podcaster, and the chairman of CoinDesk's annual conference, Consensus. He has worked as a journalist on five continents, including eighteen years with Dow Jones and *The Wall Street Journal*, and was a founding staffer at MIT's Digital Currency Initiative. Casey has five other books to his name: *Che's Afterlife*, *The Unfair Trade*, *The Age of Cryptocurrency*, *The Social Organism*, and *The Truth Machine*.